Biology, Religion, and Philosophy

The intersection of biology and religion has spawned exciting new areas of academic research that raise issues central to understanding our own humanity and the living world. In this comprehensive and accessible survey, Michael L. Peterson and Dennis R. Venema explain the engagement between biology and religion on issues related to origins, evolution, design, suffering and evil, progress and purpose, love, humanity, morality, ecology, and the nature of religion itself. Does life have a chemical origin – or must there be a divine spark? How can religious claims about divine goodness be reconciled with widespread predation, suffering, and death in the animal kingdom? Peterson and Venema develop a philosophical discussion around such controversial questions. The book situates each topic in its historical, scientific, and theological context, making it the perfect introduction for upper level undergraduates, graduate students, scholars, and interested general readers.

MICHAEL L. PETERSON is Professor of Philosophy at Asbury Theological Seminary. He is author of numerous books, including *C. S. Lewis and the Christian Worldview* (2020) and *Reason and Religious Belief* (with William Hasker, Bruce Reichenbach, and David Basinger, 2012). He is coauthor, with Michael Ruse, of *Science, Evolution, and Religion: A Debate about Atheism and Theism* (2016).

DENNIS R. VENEMA is Professor of Biology at Trinity Western University, where he specializes in molecular biology and genetics. He is coauthor, with Scot McKnight, of *Adam and the Genome: Reading Scripture after Genetic Science* (2017).

Cambridge Introductions to Philosophy and Biology

General editor
Michael Ruse, Florida State University

Other titles in the series
Derek Turner, *Paleontology: A Philosophical Introduction*
R. Paul Thompson, *Agro-Technology: A Philosophical Introduction*
Michael Ruse, *The Philosophy of Human Evolution*
Paul Griffiths and Karola Stotz, *Genetics and Philosophy: An Introduction*
Richard A. Richards, *Biological Classification: A Philosophical Introduction*
Lynn Hankinson Nelson, *Biology and Feminism: A Philosophical Introduction*

Biology, Religion, and Philosophy

An Introduction

MICHAEL L. PETERSON

Asbury Theological Seminary

DENNIS R. VENEMA

Trinity Western University

CAMBRIDGE
UNIVERSITY PRESS

University Printing House, Cambridge CB2 8BS, United Kingdom

One Liberty Plaza, 20th Floor, New York, NY 10006, USA

477 Williamstown Road, Port Melbourne, VIC 3207, Australia

314–321, 3rd Floor, Plot 3, Splendor Forum, Jasola District Centre, New Delhi – 110025, India

79 Anson Road, #06–04/06, Singapore 079906

Cambridge University Press is part of the University of Cambridge.

It furthers the University's mission by disseminating knowledge in the pursuit of education, learning, and research at the highest international levels of excellence.

www.cambridge.org
Information on this title: www.cambridge.org/9781107031487
DOI: 10.1017/9781139381765

First published 2021

A catalogue record for this publication is available from the British Library.

Library of Congress Cataloging-in-Publication Data
Names: Peterson, Michael L., 1950– author. | Venema, Dennis R., 1974– author.
Title: Biology, religion, and philosophy : an introduction / Michael Peterson, Dennis Venema.
Description: Cambridge, United Kingdom ; New York, NY, USA : Cambridge University
    Press, 2021. | Series: Cambridge introductions to philosophy and biology | Includes
    bibliographical references and index.
Identifiers: LCCN 2020039592 (print) | LCCN 2020039593 (ebook) | ISBN 9781107031487
    (hardback) | ISBN 9781107667846 (paperback) | ISBN 9781139381765 (epub)
Subjects: LCSH: Biology–Religious aspects.
Classification: LCC BL255 .P48 2021 (print) | LCC BL255 (ebook) | DDC 570.1–dc23
LC record available at https://lccn.loc.gov/2020039592
LC ebook record available at https://lccn.loc.gov/2020039593

ISBN 978-1-107-03148-7 Hardback
ISBN 978-1-107-66784-6 Paperback

For our wonderful family members ...

Our wives

Rebecca Peterson and Valerie Venema

Second-generation Petersons and their wives

Aaron and Melissa Peterson

and

Adam and Claire Peterson

Second-generation Venemas

Elijah, Davin, and Preston

Third-generation Petersons

Mason, Daisy, Brody, Ruby, and James

Contents

Acknowledgments

We would like to thank a number of persons who helped greatly in the production of this book. Susan Olin proofread the original manuscript and offered insightful advice. Mara Eller proofread during later stages of production and provided suggestions for further improvement. Chris Holland and Robert Williams, our research assistants, worked diligently on many aspects of the book, from checking the accuracy of citations to providing commentary on readability.

We would also like to thank our editor, Hilary Gaskin, who gave guidance during the book's preparation. Perhaps less visible – but still so very important to the eventual appearance of the book – is our series editor, Michael Ruse. It was Michael who had the initial vision for this kind of project and enlisted us in the service of that cause.

Last, we thank our families – wives, children, and grandchildren – who were supportive and patient during our writing, and we dedicate the book to them.

Introduction

The intersection of biology and religion has spawned an exciting academic area, attracting scholars, generating research projects, and gaining notice in general culture. Topics dealing with the relation of biology and religion are inherently interdisciplinary, making philosophy – which is also inherently interdisciplinary – essential for clarifying the issues, identifying key assumptions, and evaluating alternative positions. Therefore, in this book we develop a philosophical discussion of the major topics shaping this field of inquiry, acquainting the reader along the way with the major voices and viewpoints that have contributed to its advance.

The Table of Contents projects a panorama of important issues pertaining to biology and religion. Of course, the issues covered are located within the broader scholarship on the relationship of science and religion, which is both historical and philosophical, a relation that has been conceived in multiple ways, as we shall see. Furthermore, the biosciences are special in that they pertain to life – to the whole organic world – leading us early on to consider their relation to the sciences of the inorganic world. The subject of life – its origin, organization, and development – is a deeply vested interest for both biology and religion, raising questions about creation, evolution, design, and purpose that we must work through.

A number of other issues then readily arise. Since biology seems to reveal pervasive predation and death in the natural world, the perennial problem of evil occurs and challenges religion to square the troublesome data with ideas of a good and purposeful divine being or divine realm. Questions about progress and purpose also arise in regard to the evolutionary story – which tells of the rise of life over vast amounts of time, from humble beginnings to *Homo sapiens* – requiring response from religious perspectives. Sometimes the challenges occur more obviously within the field of biology itself, as in the curious facts

1

of love and altruism in a Darwinian world. Biology typically characterizes all life as caught in a self-interested struggle for existence, whereas all major religions teach love, kindness, and even self-sacrifice. Particularly among humans, whose capacity for ostensible other-regarding behavior seems greater than that of all other animals, the encounter of biology and religion raises the question of whether genuine altruism and love are possible.

Some issues that we treat in the book arise because an evolutionary account of key phenomena can seem incompatible with a religious explanation. Traditionally, religious accounts maintain that ethics is a form of objective knowledge based on universal moral truths, given by deity, that people are obligated to obey. However, some thinkers have construed the evolutionary shaping of human moral beliefs and emotions to imply that ethics has no objective status, religious or otherwise. Similarly, most major religions see themselves as connected to a divine source and as teaching true beliefs and right ways of living. However, biological theories of religion claim to explain religion in evolutionary terms and thus to preempt any question of a higher or supernatural dimension. Another issue on which evolutionary and religious perspectives interface is the environment. Part of our ever-increasing evolutionary knowledge is greater understanding of ecology, of how organisms are related in interconnected systems with which we humans interact. Increased knowledge of ecology has prompted considerable religious reflection on the environment.

As an introductory text, this book is composed of chapters on topics that themselves could generate multiple more specialized monographs. However, since we present a broad treatment – covering many topics that are somewhat eclectic, engaging the relevant but diverse biological subdisciplines, and seeking to weave various religions into the discussion – we cannot delve deeply into any topic in technical detail. Instead, we offer a narrative that exposes the reader to a fascinating array of biological and religious questions in interaction, while providing historical, scientific, and theological context as preliminary to philosophical appraisal. Among the readers this volume targets are upper-division undergraduates, graduate students, scholars wanting a survey of the issues, and interested persons in the general public. While the issues addressed are of intrinsic interest within the fields of biology, religion, and philosophy, these issues also have social importance because they continue to influence our human self-understanding and even affect some matters of public policy.

Courses in which this book would be an ideal core text, or even an ancillary text, include introductory philosophy of biology courses and philosophy of religion courses; it would also fit well with religious studies, theology of nature, or theological anthropology courses. For purposes of learning or for teaching, this book unfolds an expansive conversation that presents difficult issues in understandable terms, endeavors to be fair and balanced to all sides, and yet indicates at certain points the direction in which a more likely solution might be sought. Throughout our discussion, we show how, in many cases, it is so often the philosophical interpretation of biological facts or religious claims that is really at stake. The philosophical clarity we aim at will help the reader adopt a proper reflective posture toward the issues, transcending their complexities in order to navigate them intelligently.

Part I

General Issues

1   Science, Biology, and Religion

A discussion of the relationship between biology and religion is a subset of the larger conversation about the relationship between science and religion. At the turn of the seventeenth century, the new astronomy catalyzed the Scientific Revolution, which raised questions about whether the Roman Catholic Church or practicing scientists were properly entitled to make claims about the structure and operation of the heavens. As modernity unfolded, the Newtonian Revolution consolidated its position: the purview of the natural sciences – from astronomy to physics to chemistry – was to make claims about the structure and operation of the physical world that were grounded in empirical research and not religious dogma. In the middle of the nineteenth century, the science of biology, mostly in the form of natural history and practiced by Darwin and others, caused new tensions with religion. Biology has remained at the center of much controversy with religion – and now, given their influence in contemporary life, what we may call the emerging "biosciences" present new challenges to which religion must continually respond.

The intersection between biology and religion requires careful analysis because the historical engagement between these two important human activities has been complex and because these activities remain deeply enmeshed in cultural and political power structures. However, our philosophical approach offers a way forward. Philosophy helps us identify ontological and epistemological commitments, clarify alternative positions, define terminology, analyze connections across areas of knowledge and human experience, and evaluate arguments on important issues. To begin our exploration of the issues, we clarify the nature of religion and the nature of science, respectively, and then survey the major typologies or models reflecting the different ways their relationship has been conceived. This philosophical work is a prelude to clarifying the particular nature of biology as a science so that we may fruitfully study its interactions with religion. The foundational and integrative role of philosophy plays out as we interact with a variety of positions on the topics under study. Our dialogical format thus provides a context and sets the stage for the reader to navigate through the pertinent issues.

Religion in Human Life

Arriving at a precise definition of religion is notoriously difficult, and yet attempts at definition abound. C. P. Tiele writes that "[r]eligion is, in truth, that pure and reverential disposition or frame of mind which we call piety." "Religion," claims James Martineau, "is the belief in an ever living God, that is, in a Divine Mind and Will ruling the Universe and holding moral relations with mankind." F. H. Bradley states that "[r]eligion is rather the attempt to express the complete reality of goodness through every aspect of our being." Each of these definitions keys on some characteristic associated with religion: Tiele accents the attitude of piety; Bradley links religion with goodness; and Martineau features belief in ethical monotheism. Other definitions touch upon traits such as ritualistic acts, prayer and communication with gods, and so on.[1]

Wisdom counsels us, however, not to define religion by elevating any single feature to the status of universal definition because such treatments admit of counterexample. Tiele's definition is incomplete because shamanistic religions, for example, do not involve feelings of genuine piety so much as they promote prudential or utilitarian acts of obeisance. Likewise, Martineau's definition fails to cover ancient polytheistic religions (such as those of ancient Egypt and Greece) that do not recognize a single divine moral ruler of the universe. Frankly, these and other readily available counterexamples show that defining religion too generically distorts the rich, complicated particularities of each religion. Even attempts to specify the most general aspects of religion, say, by connecting religion with the idea of worship or with the need for the divine or the supernatural, are too narrow. The notion of a supernatural realm does not even occur, for

---

[1] W. P. Alston, "Religion," in P. Edwards (ed.), *Encyclopedia of Philosophy*, 8 vols. (New York: Macmillan, 1967), vol. VII, 140.

example, in the nontheistic schools of Buddhism, which seek Nirvana, and it functions in very different ways, say, in Taoism and Hinduism. The great differences among religions make it extremely difficult to find a common denominator or to talk about "religion" in the abstract.

Instead of offering a universal definition that is subject to counterexample, religion scholar Ninian Smart suggests that we start by identifying the common dimensions of all religions; he designates seven.[2] First, the "doctrinal dimension" involves the accepted beliefs – perhaps few and unsystematic or perhaps many and highly organized – about ultimate reality or the divine and its relation to humanity. Each religion, second, also has a "mythological dimension" that conveys its particular understanding of the religious ultimate to faithful adherents in terms of symbolic speech and stories. Third, certain moral actions and general life orientations are associated with what it means to embrace and follow a given religion: this aspect is the "ethical dimension." Fourth, the "ritual dimension" pertains to the prescribed behaviors, both public and private, that are thought to reflect worship of the divine or properly relating to the ultimate. Fifth, the "experiential dimension" of religion, both personal and collective, reflects what it is like to act and live as a religious believer. The experience can range from a quiet sense of the presence of a god in daily life to the highly mystical consciousness of union with ultimate reality. Sixth, the "social dimension" is how a religion organizes all sorts of interpersonal relationships. Last and seventh, the "material dimension" of a religion pertains to how the gods or god or ultimate religious reality is reflected in the physical world. The material dimension can be simply how the divine is conceived in relation to the world (say, as the ancient Greek god Poseidon is associated with the ocean) or how a religious community designs art and architecture to create an atmosphere of worship (say, in the great Christian cathedrals of Europe).

Now, after recognizing the difficulty of identifying one trait that defines religion and after appreciating the value of characterizing religion by its key dimensions, we, nevertheless, venture a working definition for the purposes of our developing discussion. Let us say that "religion" is *a human phenomenon that is constituted by a set of beliefs, actions, and experiences, both personal and collective, organized around the concept of an ultimate reality that inspires or requires*

[2] N. Smart, *Dimensions of the Sacred* (Los Angeles: University of California Press, 1999).

a certain response like devotion, worship, or focused life orientation.[3] This reality may be understood as a unity or a plurality, personal or nonpersonal, divine or not, differing from religion to religion. Yet, it seems that every cultural phenomenon that we call a religion fits this definition. The prescribed actions vary from ritualistic patterns of behavior to general ethical living; the desired emotions vary from feelings of piety and humility to a sense of optimism about life and the universe.

For the most part, our study of the relation of science and religion revolves around the recognition that all religions have *beliefs* – that is, a doctrinal dimension, as Smart would label it – whether the beliefs are rather simple or sophisticated. Our inquiry into the relationship between science and religion generally, or into the relationship between biology and religion more specifically, takes religious beliefs seriously. In philosophy, we typically say that a belief is propositional, that it is expressible in terms of an assertion that can be true or false, probable or improbable, and the like. Specifically, *religious* beliefs relate in one way or another to what a religion teaches about reality, including knowledge, morality, humanity, and a number of other key features of life and the world. In a certain sense, then, every religion rests on a set of beliefs that function conceptually as its worldview core, but this core also undergirds and gives sense to specifically religious observance as well as daily living.

Science as a human activity also generates beliefs – which are, again, claims about the world and human life, assertions that can be understood and discussed and can be true or false; therefore, a crucial area of inquiry into the relation of science and religion obviously pertains to the respective beliefs they hold, their grounds and implications. Of course, at another level, science also offers theories or explanations of religion as a human phenomenon – psychological, social, and biological explanations of religion that have some measure of theoretical and empirical support.

As we develop our exploration of the relation of science, and particularly biology, to religion, we will often transition from discussing religion in general to discussing theism in particular because of its important role in Western culture. Theism asserts that there is an omnipotent, omniscient, perfectly good being that created, sustains, and interacts with the world and

---

[3] M. L. Peterson et al., *Reason and Religious Belief: An Introduction to the Philosophy of Religion,* 5th ed. (New York: Oxford University Press, 2013), 7.

all it contains. Although theism is not a living religion, it is the common conceptual core of the three great Abrahamic religions: Judaism, Christianity, and Islam. Historically, in the West, theistic ideas have interacted in a variety of ways with science and thus require specific attention in our treatment of religion and the biosciences.

Science in Human Life

Science is one of the most impressive knowledge-gathering projects in human history, providing an astounding amount of information about the world and promising much more. Like religion, science is an important human activity, shaping so much of our world and exerting enormous influence. To begin our discussion of science, we face the same question of definition that we did with religion. What is the definition of science? How shall we characterize it as a human activity in its own right? Let us begin our discussion at a very basic level, observing science both as a way of seeking knowledge and as a body of accumulated knowledge.

The method of science is at the heart of its success in gaining insight into the workings of the physical world. However, the scientific method is not a single procedure but a number of practices in which scientists engage, from observing and experimenting to creative hypothesizing and constructing models. The key is the intentional rigor of scientists in tying hypotheses to empirical experience through experimentation. Hypotheses are used to make predictions, which are then tested. Those hypotheses that are experimentally supported are provisionally retained and used to form additional predictions. Over time, a hypothesis might become so well supported that it becomes viewed as a theory; in the scientific sense, a theory is a broad explanatory framework that makes sense of a large swath of experimental data and has not yet been falsified or overturned. Theories may be modified in light of new evidence or even, in principle, discarded if shown to be inadequate. Thus, a theory is a conceptual tool accepted by scientists as enjoying a high measure of corroboration, and some theories are even felt to be so well supported that new evidence is unlikely to substantially modify them. Science is an important expression of the human drive to understand the physical universe – how it is structured and how it works – and it remains the most productive method to that end that humans have thus far conceived.

We typically say in science that we have an understanding of a given phenomenon when we have a well-supported theory. The explanatory work of science is, in turn, anchored in its ability to identify causes of the phenomena under study. To do its work, science must assume that there are causes – that is, physically necessary connections between events – and that these causal connections can be codified as scientific laws. Scientific explanation, then, brings empirical phenomena under known laws and explains them by means of theories. Although this basic characterization of science might seem uncontroversial now, it was born out of historical controversy. Ancient science, inspired largely by Aristotle, was a priori and nonexperimental in character because it sought understanding of the behavior of any given physical phenomenon by means of pure insight into its essential nature. However, at the beginning of the seventeenth century, this way of doing science was rejected, effectively giving birth to modern science. Now, the scientist first formulates a hypothesis about what causes a particular object to operate in certain ways and then tests that hypothesis empirically through observation and experiment.

Of course, even in early modernity, there were misinterpretations of the new scientific procedure. Francis Bacon gave rise to the famous misunderstanding that science begins by collecting data, from which it draws conclusions through inductive reasoning, a view that, unfortunately, is still taught in high school science books and believed in general culture. This narrow inductivist view of the scientific method makes it seem like one does science just by putting on a white coat on a Monday morning and walking into the lab and gathering data. Instead, the scientist begins with questions about a phenomenon and then formulates a hypothesis about why it is the way it is or functions as it does. Only then is the experiment designed and data collected for evaluation.

In a book covering the biosciences, it is fitting to quote Charles Darwin's rejection of the Baconian view of science. In a well-known passage in his correspondence, Darwin precisely identified the error of narrow inductivism:

> How profoundly ignorant [Bacon] must be of the very soul of observation! About thirty years ago there was much talk that geologists ought only to observe and not theorize; and I well remember some one saying that at this rate a man might as well go into a gravel-pit and count the pebbles and

describe the colors. How odd it is that anyone should not see that all observation must be for or against some view if it is to be of any service![4]

The Baconian definition simply fails to see that the scientific method does not begin with data but with problems and puzzles about the behavior of physical phenomena that we do not fully understand. Nothing is recognized as data in science unless it is related first to a hypothesis, which, if empirically supported, advances our understanding. Thus, it is the careful construction of testable hypotheses and, subsequently, the highly structured attempts to test those hypotheses that form the essence of science.

Note that the underlying ontological assumption of science mentioned earlier – that there is a physical world structured by causal regularities that can be codified as laws – is now obviously coupled with the key epistemological assumption of science: that human beings have the capacity to access these regularities. The human capacity to explain the physical world according to causal laws finds sophisticated expression in the scientific method. In the philosophy of science, this characterization of the ontological and epistemological assumptions describes a realist view of science. Philosophical realism broadly holds that there is a real world independent of our minds and that we have the cognitive ability to access it, recognizing, nonetheless, that our knowledge of the world is revisable and not perfect and that knowledge is mediated through our own cognitive structures and our social situation. However, the dual confidence remains that there is a regular world and that our knowledge of it is objective. In fact, the growth of scientific knowledge in early modernity gave rise to the belief, rooted in Aristotle, that the world is a physical system, characterized by a total, coherent set of laws, and that our knowledge of it can progressively increase.

Although the unfolding discussion of this book interacts largely with a realist view of science, it is helpful to note before proceeding that various nonrealist or antirealist views of science have been proposed in the history of the philosophy of science. Instrumentalism, for example, holds that a successful scientific theory does not reveal anything about the structure, properties, or processes of nature itself but instead provides a summary of the behavior of a natural phenomenon and a valuable predictive tool for its future behavior. Thus, for instrumentalists, the question of whether a

---

[4] Darwin to H. Fawcett, September 18, 1861, Letter no. 3257, Darwin Correspondence Project, www.darwinproject.ac.uk/letter/DCP-LETT-3257.xml.

scientific law is actually true about some aspect of physical nature is side-stepped in favor of using the law as a predictor. Instrumentalism in particle physics, for example, avoids the question of whether a particle is a discrete entity with individual existence or is rather the excitation mode of a certain region of a field. Instead, it focuses on the usefulness of the theoretical term "particle" to predict outcomes. We note that the practice of science – pertaining to how we can get along doing science – is different from the fundamental philosophical question of how effective practice is best explained in ontological and epistemological terms.

Probably the most famous version of scientific antirealism was advanced by Thomas Kuhn in *The Structure of Scientific Revolutions*. In this 1962 book, Kuhn describes the working of science as a communal activity of researchers who operate according to the prevailing "paradigm" – a shared understanding of the physical world and the yet-unsolved problems about it they investigate. Essentially, paradigms are conceptual frameworks, shared by a community of inquirers, that contain the solved problems and project the unsolved problems of the science in question, providing implicit directions and limits to theorizing and research.[5] At one point, Kuhn explains that paradigm thinking is somewhat like metaphorical thinking, calling attention to the fact that scientific thought and language are deeply metaphorical and that scientific knowledge is, in the end, socially constructed, thus sparking criticism that his theory does not adequately account for objectivity in science. Kuhn's work may be seen as part of the movement in epistemology holding that knowledge is "socially constructed" and is not a pure, pristine representation of objective reality.

Although our developing discussion assumes a generally realist view of science, it is helpful to note the idea that, in any period of time, science interacts with its culture to give rise to what we may call the scientific picture of the world. Philosopher of biology Michael Ruse states that this picture is in a real sense a metaphor – a word or figure of speech applied to the world that is not literally applicable. "We look at the world, or parts of it," Ruse states, "through the lens of something with which we are familiar, spurring us to ask questions and (with luck) to find answers."[6] He adds that

---

[5] T. Kuhn, *The Structure of Scientific Revolutions*, 2nd edn. (Chicago: University of Chicago Press, 1970).

[6] M. L. Peterson and M. Ruse, *Science, Evolution, and Religion: A Debate about Atheism and Theism* (New York: Oxford University Press, 2016), 29.

the particular sciences are "drenched" in their own metaphors – force, work, attraction, genetic code, natural selection, plate tectonics, Oedipus complex, and more. Work in cognitive linguistics reveals that metaphorical thinking in science is a reflection of the broader practice of metaphorical thinking in ordinary life as we seek to organize our experience.[7]

One way of characterizing the historic tension that occurred between science and religion is to say that the most dominant metaphors – the root metaphors – of science changed from ancient Aristotelian science, which viewed nature as an organism, to modern science, which views nature as a machine. From the time of the Greeks, nature was seen as a self-contained organism and studied in organic terms. However, classical Christianity taught that nature was not a self-contained whole but a divine artifact, a creature made by a supreme being and endowed with laws and harmonies. By successive steps, over many centuries, the Christian understanding of nature led to the idea that nature was "a divinely organized machine in which was transacted the unique drama of Fall and Redemption."[8] Ironically, since God was spirit, he was eventually seen as removed from the universe, which was completely material. At a very fundamental level, the mode of scientific explanation had to change to fit the shift in metaphor. If a self-contained nature is imbued with purpose, then purposive or teleological explanations for the behavior of natural objects were appropriate – indicating *why* something does what it does. However, if nature is machine-like, then mechanical explanations were appropriate – indicating *how* some material thing works the way it does.

The Scientific Revolution – that great transformation of our understanding of the natural world that began with Copernicus in the middle of the sixteenth century and ended with Isaac Newton at the end of the seventeenth century – changed our metaphors about nature and our ways of explaining it. In the Galileo affair, which was the symbolic birth of modern science, the tension between religion and science was partly over a conflict in metaphors. On the one hand, the Catholic Church insisted that the geocentric theory of Ptolemy provided the true picture of the cosmos, a position supported by the Church's teleological view that humans are the center of God's concern.

---

[7] G. Lackoff and M. Johnson, *Metaphors We Live By* (Chicago: University of Chicago Press, 1980).

[8] A. R. Hall, *The Scientific Revolution 1500–1800* (London: Longmans, Green, 1954), xvi–xvii.

When Galileo strongly supported the heliocentric theory of Copernicus, his view came into direct conflict with the Church. Although Galileo, a faithful believer, was essentially applying the Christian idea that the divinely created material cosmos could be studied empirically and explained mechanically, the Church insisted that its particular teleological explanation determined what mechanical explanation was acceptable, a posture that was eventually shown to be indefensible. As the Enlightenment progressed in Europe, then, the burgeoning sciences prospered as they relinquished teleological explanation and developed a mechanistic model of explanation, with new empirical theories, some of which were so well supported that they became scientific laws.

By the end of the eighteenth century, all of the sciences were mechanistic except biology, the science of the living world. Immanuel Kant thought it impossible that there could be a mechanistic explanation of organisms because they seem purposively constructed: "We can boldly say that it would be absurd for humans even to make such an attempt or to hope that there may yet arise a Newton who could make comprehensible even the generation of a blade of grass according to natural laws that no intention has ordered; rather, we must absolutely deny this insight to human beings."[9] Later in the nineteenth century, however, the work of Charles Darwin would revolutionize biology with the recognition that all organisms are the end products of a long, slow process of adaptive change. In *On the Origin of Species*, published in 1859, Darwin argued that species arose from common ancestors as natural selection acted on heritable variation. Thus, Darwin's work showed that there was a lawlike mechanism that accounted for organic structure and function, an insight that revolutionized biology and brought that science under the machine model with the rest of the sciences. After Darwin, there was scientifically no apparent need for teleological explanation in science, but not all religious people agreed, a point we trace in several discussions later in the book.

Conflict or Compartmentalization?

Our previous discussion of the respective natures of science and religion now serves as prelude to further exploration of how to think about their

[9] I. Kant, *The Critique of Judgment* (New York: Hafner Publishing Company, 1951), 270.

relationship. Ian Barbour has identified four major conceptions – which he calls "models" – of the science–religion relation: conflict, independence, dialogue, and integration.[10] Referring to the natural sciences to define these models, let us consider each one in turn.

Perhaps the most dominant image of the science–religion relationship in contemporary culture is one of irresolvable conflict. The conflict or "warfare" model has deep roots, as old as the Galileo affair and as contemporary as the creation/evolution controversies in America. Other tensions between science and religion abound, from relativity theory in physics – which challenges religious perspectives on God's relation to the world due to changing concepts of space, time, and causality – to artificial intelligence research – which calls into question the unique status of human beings as we find that computers can perform increasingly complex reasoning calculations and even "learn" new things.

Early in the twentieth century, two dramatically opposed schools of thought greatly solidified the cultural idea of inherent conflict: scientific materialism and biblical literalism. Scientific materialism was very much shaped by logical positivism, which held that all intellectually serious beliefs must be verifiable or falsifiable by empirical experience, giving rise to the epistemological view that the method of science is the only reliable procedure for obtaining knowledge. Religious beliefs, then, cannot be knowledge because they seem private and parochial. This epistemological view dovetails with the metaphysical view that the physical world that science studies is the sum total of reality. Thus, there is no supernatural reality – involving God, soul, or afterlife – to which religion can meaningfully relate. Although the term "scientific materialism" is no longer used, its basic epistemological and metaphysical assumptions underlie many current approaches to the science–religion relationship. A group of thinkers who are often labeled the "New Atheists" have become famous for defending this same basic viewpoint in contemporary culture. Richard Dawkins, Daniel Dennett, Sam Harris, and Christopher Hitchens are particularly prominent representatives of the view that science discredits religion and supports philosophical naturalism and materialism.

---

[10] I. G. Barbour, *Religion and Science: Historical and Contemporary Issues* (New York: HarperOne, 1997), 77–105.

Biblical literalism, at the other end of the spectrum, is a distinctively American phenomenon within Christianity that began around the turn of the twentieth century. During the late nineteenth and early twentieth centuries, "Protestant fundamentalism" – or simply "fundamentalism" – began as a reaction to several movements in intellectual culture: the rise of Darwinian biology, Freudian psychology, and German "higher criticism" of the Bible.[11] Tenaciously insisting on a literal interpretation of the Bible, fundamentalists taught that the book of Genesis indicates that God created the universe in six literal twenty-four-hour days and instantaneously created humanity at the end of the sixth day. Coupled with the calculation based on biblical texts that Earth is about 6,000 to 10,000 years old, this fundamentalist outlook clashed dramatically with the scientific claims that the planet formed about 4.5 billion years ago, that life developed within the first billion years, and that *Homo sapiens* appeared only after untold millions of years of evolutionary development.

From the perspective of 2,000 years of Christian thought, which contains a variety of views on the relation of scripture and science, biblical literalism can be seen to be an anomaly. St. Augustine, for example, famously maintained that when some particular passage of the Bible appears to conflict with established facts or scientific information, that passage should probably be reinterpreted, perhaps figuratively.[12] Other Christian medieval thinkers acknowledged that the Bible includes a rich diversity of literary genres, reveals truth at many levels, and was never meant to be a scientific document. Interestingly, in 1983 Pope John Paul II articulated a stance quite different from the one the Roman Catholic Church took toward the Galileo controversy when he asserted that we now have "a more accurate appreciation of the methods proper to the different orders of knowledge," thus clearly giving place to both religious and scientific knowledge.[13] Nevertheless, the conflict model was dominant through the twentieth century and is still dominant at the beginning of the twenty-first century. This has been particularly true in America, where science has often been understood in strict empiricist terms and Christian belief has often been

---

[11] R. E. Olson, *The Story of Christian Theology* (Downers Grove: IVP Academic, 1999), 554–569.
[12] Augustine, *The Literal Meaning of Genesis*, 1.19.
[13] John Paul II, "Address on the occasion of the 350th anniversary of Galileo's publication," *L'Osservatore Romano*, English weekly edn. (May 30, 1983), 7. See his encyclical *Fides et Ratio* (1998). Available online at www.vatican.va.

aggressively projected in fundamentalist terms. The conflict in various forms still lingers, as we will see in several chapters to follow.

However, some more recent thinkers have been interested in a position more moderate than one of outright conflict, proposing ways of thinking about science and religion that make them completely independent activities that cannot in principle be at odds. The "independence model" asserts that each field has its own distinctive function in human life, making for separation or compartmentalization. Various perspectives on the religious side have supported the view that science and religion are independent. Religious existentialism, originating with Søren Kierkegaard in the nineteenth century, insists that the heart of religion is the risky choice to live authentically in pursuing religious values as the basis of meaning in one's life. Also, Protestant neoorthodoxy in the twentieth century specifically emphasized the primacy of "special revelation" – God's self-disclosure in scripture – as the sole source of religious knowledge.[14] Karl Barth, the most famous representative of neoorthodoxy, argued that religious knowledge is self-authenticating, carrying its own validation. To him, religious knowledge was not dependent on natural theology, which sometimes utilizes knowledge from science. Continuing the existentialist theme, noted Jewish theologian Martin Buber stressed that the individual's relation to God is an "I-Thou" relationship, deeply personal and subjective. In contrast, science studies nonpersonal objects through relationships he characterized instead as "I-It."[15]

The independence model has its advocates on the science side as well. Scientist Stephen Jay Gould used the Roman Catholic idea of a "magisterium," which is a domain of teaching authority, to support the model. For Gould, science and religion are "nonoverlapping magisteria" – entirely separate domains over which each discipline has teaching its respective authority. He advances this idea as a principle that he dubs by the acronym NOMA – nonoverlapping magisteria – and explains as follows:

> The net of science covers the empirical realm: what is the universe made of
> (fact) and why does it work this way (theory). The net of religion extends over
> questions of moral meaning and value. These two magesteria do not overlap,

[14] Olson, *Story of Christian Theology*, 570–589.

[15] M. Buber, *I and Thou*, trans. W. Kaufmann (New York: Charles Scribner's Sons, 1970).

nor do they encompass all inquiry (consider, for starters, the magisterium of art and the meaning of beauty).[16]

In eliminating conflict between science and religion, the independence model conceived by Gould claims the realm of fact for science, which he considered objective, but cedes the realm of value and meaning to religion, which he considered subjective. Gould concluded that the independence model has a double effect. It prevents religion from dictating to science and prohibits science from claiming higher moral or intellectual insight than religion, all of which may inspire mutual humility and provide the basis for a larger vision of reality.

Is Dialogue Possible?

The dialogue model seeks to go beyond conflict or compartmentalization and foster mutual understanding between science and religion by seeking common ground. Since the 1990s, conscious pursuit of this model has greatly enhanced contact between science and religion. In recent decades, the pursuit of this model has borne fruit. The John Templeton Foundation has been particularly involved in funding conferences and scholarly research projects, both of which have resulted in a number of books and articles on the topic. Let us consider two avenues proposed by Ian Barbour along which science and religion might have meaningful dialogue: boundary questions and methodological parallels.

Boundary questions delve into how science points beyond itself. Philosophers of science point out that science rests on certain *presuppositions*, assumed beliefs that shape its whole enterprise. One key boundary question asks where the presuppositions required for the foundations of science come from, since science cannot establish them by its own methods. These presuppositions are essential to provide a sketch of the fundamental characteristics of the natural world that science investigates and of the capacities of human beings as scientists to investigate and know about it. For example, science must assume, but cannot by its own methods validate, the belief that nature is physical, real, and accessible to rational investigation. Furthermore, in order to do its work, science must also assume that human beings have the rational

---

[16] S. J. Gould, "Two Separate Domains," in M. L. Peterson et al., *Philosophy of Religion: Selected Readings*, 5th edn. (New York: Oxford University Press, 2014), 541.

capacity to investigate nature and learn about its inherent lawlike operations. It does no good to say that science investigates our human powers of inquiry, because those powers must already be trusted in order to begin that investigation.

These beliefs that are foundational for science are drawn from a vastly different philosophical worldview. The belief that nature is real and rational may seem unremarkable, but it was not assumed by ancient Greek science, which had to be supplanted for modern science to be born. For the Greeks, the material world was both less real than the world of ideas and also inherently disordered, which is why Greek science specified a priori how things necessarily must behave rather than engaging in disciplined inductive empirical investigation. Modern science arose in the context of yet another philosophy of nature, one rooted in the Judeo-Christian doctrine of creation. E. L. Mascall argues that the worldview serving as the intellectual backdrop for the pioneers of early modern science was radically different from the intellectual backdrop of Greek science:

> A world which is created by the Christian God will be both contingent and orderly. It will embody regularities and patterns, since its Maker is rational, but the particular regularities and patterns which it will embody cannot be predicted a priori, since he is free; they can be discovered only by examination. The world, as Christian theism conceives it, is thus an ideal field for the application of scientific method, with its twin techniques of observation and experiment.[17]

In the same vein, Mascall argues for another assumption drawn from the classical Judeo-Christian doctrine of creation: that the human rational ability to know physical nature enables robust empirical inquiry.

Another boundary issue arises when science reaches the limit of its abilities to explain an important phenomenon. "Big Bang" science is a perfect example, because, in pushing back to the earliest event in the cosmos, astronomers and theoretical physicists ask questions about the prior conditions that precipitated that initial cataclysmic event. Physicist Stephen Hawking, using M-theory, has explained that the initial singularity from which the universe expanded indicates that the operation of quantum gravity got everything going and, by implication, makes creative activity by God

---

[17] E. L. Mascall, *Christian Theology and Natural Science* (New York: Ronald Press, 1956), 132.

unnecessary.[18] Since considering ultimate beginnings takes us as far back in time as science can extend its explanatory reach in terms of known laws, philosophers say that science has reached a limit or boundary and thus open a path for dialogue between theology and science.

At this level of inquiry, profound questions arise that are not clearly just scientific – such as, why is there a law such as gravity, which Hawking thinks created everything else? Or, better, why is there something rather than nothing; why the law of gravity? How does reference to the abstract law explain the existence of a concrete universe? Theologian and physicist John Polkinghorne indicates that these kinds of questions are at the limits of science, for they open the door for metaphysics and theology to say something about God as the creative ground of the existence and lawful order of the universe.[19] Of course, some nonreligious thinkers counter by proposing that the ordered cosmos arose by pure chance rather than by divine activity. One move they make on the side of pure chance is to argue that there is an infinity of possible universes out of which one universe will be actual, such that it is by chance that our specific universe exists, albeit a very fortunate occurrence that brought about our lawlike and ordered universe.[20] The debate in effect comes down to whether the ultimate, rock-bottom explanation for the existence of this universe is mechanistic or teleological. Unless we claim for science the supremacy and exclusiveness of reductionist empiricism and materialism, boundary questions invite fruitful science–religion dialogue.

We turn now from the topic of boundary questions to a discussion of methodological parallels between religion and science. Two productive topics for exploring methodological parallels pertain to the role of paradigms and to the nature of "research programmes." Barbour believes that the idea of a paradigm provides insight into the ways in which both science and religion operate. On the science side, some philosophers of science, such as Norwood Russell Hanson and Stephen Toulmin, have argued that science is not pristine, objective, and free of bias, as popular stereotypes suggest. They

[18] S. Hawking and L. Mlodinow, *The Grand Design* (New York: Bantam Books, 2010), 180–181.
[19] J. Polkinghorne, *Science and Theology: An Introduction* (Minneapolis: Fortress Press, 1998), 79–81.
[20] V. Stegner, *God and the Multi-verse: Humanity's Expanding View of the Cosmos* (Buffalo, NY: Prometheus Press, 2014).

note a dimension of "personal involvement" in science, which makes it subjective in a sense – or, better, intersubjective and communal. No philosopher of science, however, has had more impact on these discussions than Thomas Kuhn, who argued that theory selection depends on the prevailing paradigm of the scientific community, which is historically conditioned and value-laden in various ways. He describes a paradigm as functioning in two basic ways:

> On the one hand, [the term "paradigm"] stands for the entire constellation of beliefs, values, techniques, and so on shared by the members of a given community. On the other, it denotes one sort of element in that constellation, the concrete puzzle-solutions which, employed as models or examples, can replace explicit rules as a basis for the solution of the remaining puzzles of normal science.[21]

"Normal science" simply means the efforts of the scientific community to solve the research problems they face according to the prevailing paradigm.

The paradigm contains examples of puzzles already solved and helps decide what could count as an adequate solution to other puzzles. An established paradigm is resistant to simple falsification by a few negative instances and can often be preserved by arguing that these instances are anomalies or by articulating ad hoc hypotheses. Theology may be seen as operating by a widely accepted paradigm that is then used to address questions and problems that arise within its scope, such as the problem of evil.

Kuhn's idea of a "scientific revolution" is also highly suggestive of parallels with religion. Normal science, which is typically conservative and controlled by tradition, enters crisis when the long-accepted paradigm encounters increasing difficulty solving some important puzzles. At some point, the scientific community becomes dissatisfied and is attracted to an alternative paradigm because of its ability to account for existing data while handling new data in a more helpful way. When such conditions are present, according to Kuhn, science undergoes a major "paradigm shift," which is a "scientific revolution." The shifts prompted by Copernicus and Mendel would be examples of scientific revolutions. Similarly, when the accepted theological paradigm – the dominant way of looking at the world and explaining and responding to important life situations – comes under

---

[21] Kuhn, "Postscript – 1969," in *Scientific Revolutions*, 2nd ed., 175.

pressure, and a new paradigm seems promising to a significant number of influential thinkers in the religious community, a "theological revolution" is almost inevitably brewing. One could interpret, say, the Protestant Reformation in breaking from Roman Catholicism or the modern reformulation of traditional Christian doctrines by feminist theologians in this light. We could also see the emergence of Mahayana Buddhism from Theravada Buddhism, for example, as a major paradigm shift.

All of this highlights the social and communal nature of paradigms in conditioning what we call knowledge. Although debates continue about the exact degree of subjectivity in science, Kuhn's provocative analysis has inspired some to suggest that religious traditions can also be viewed as communities sharing a common paradigm. For religious communities, relevant data would be religious experience, historical events, sacred texts, and so forth – all interpreted and given significance within the paradigm. Challenges to religious belief, like challenges to a scientific theory, can be deflected by calling them anomalies or by proposing ad hoc hypotheses. Thus, the tendency of religious believers to maintain their beliefs even in light of seemingly contrary evidence is not drastically different from the behavior of scientists working under their own shared paradigm.

For those who think that a paradigm interpretation is an extreme view of the nature of theories in both science and religion, philosopher of science Nancey Murphy suggests that it is better to interpret each field as operating according to a research program that guides inquiry.[22] She draws her idea from the work of philosopher of science Imre Lakatos, who argued that a scientific community engages in ongoing projects that in one way or another preserve an accepted core theory that is supported by auxiliary theories. Thus, in light of difficult data, it is the auxiliary theories that may be modified or rejected in order to keep the difficult data from overturning the core theory. For Lakatos, viewing science – or various areas of science – as following a research program explains the tendency of scientists to cling to their main theory in light of seemingly adverse data, while at the same time accounting for their ability to make appropriate theoretical adjustments. For example, when the behavior of the perihelion of Mercury was found anomalous with respect to Newtonian mechanics, that did not in itself precipitate

---

[22] N. Murphy, *Theology in the Age of Scientific Reasoning* (Ithaca NY: Cornell University Press, 1990).

the collective abandonment of classical physics but rather stimulated auxiliary theories about the phenomenon.[23] Of course, in this case, Einsteinian physics eventually superseded Newtonian physics.

Similarly, Murphy argues that theology as an intellectual discipline proceeds by extending the scope of its core theory and defending it against difficult data with auxiliary hypotheses when necessary. The core of the Christian theological research program would contain the theologian's judgment about how to sum up the essential minimum content of the faith community – perhaps revolving around the loving and holy nature of God and God's revelation in Jesus. The next step would be to develop auxiliary hypotheses to be explained by the core and whose future modification could help protect the core. The last step, if theology is to be genuinely parallel to science, is for the theologian to seek data that help confirm the core theory and the auxiliary hypotheses related to God's goodness. For example, the positive data would include a range of religious experiences (such as a sense of providence, joy, and communal support). However, potentially negative data might be the evil and suffering in the world, which can be taken as evidence that there is no good God. Instead of surrendering the core theory, which is the theological foundation, theodicies about why God allows evil can be formulated to protect the core theory (such as theories about character building or strengthening faith). In the end, whereas a Kuhnian approach takes both science and religion in a subjective direction, Murphy sees both science and religion as having an objective quality – developing theories that have to be accountable to both all of the data and intersubjective testing by the community of inquirers.

Attempts at Integration

Although a dialogue model looks more promising than conflict or compartmentalization, some thinkers press further toward a more organic relationship between science and religion. The integration model is grounded in the intellectual ideal that human beings should seek a comprehensive and unified understanding of reality. After all, if reality is rationally coherent, and if truth is self-consistent, then surely science and theology must somehow be

[23] I. Lakatos, *The Methodology of Scientific Research Programmes, Philosophical Papers*, 2 vols. (Cambridge: Cambridge University Press, 1978), vol. I, 8–101.

harmonious. However, there are different versions of integration between the content of science and the content of theology, as identified by Barbour: natural theology, theology of nature, and systematic synthesis. In effect, each involves a different approach to how the content of science and the content of theology can be combined.

Traditional natural theology reasons from the existence of the cosmos itself or from some particular feature of the cosmos – such as its general order or human moral awareness – to the existence of God. Rather than rely on sacred revelation or church authority, natural theologians construct arguments based on human reasoning from some observable fact. Perhaps the most popular piece of natural theology is the teleological argument, which historically took several forms, and which will become relevant to later discussions in this book. Historically, Isaac Newton, Robert Boyle, and other early scientists extolled the evidences of design in nature, while William Paley in the same vein articulated a famous rendition of the argument from design. Of course, the great skeptical philosopher David Hume effectively critiqued the design argument in the eighteenth century, and Charles Darwin's work in the nineteenth century put further pressure on the design argument. Chapter 3 thoroughly addresses this important controversy, which continues to draw much interest today.

A different kind of integration of science and religion is represented in what we may call a theology of nature. Unlike those doing natural theology, those engaged in forming a theology of nature do not start from science and then construct an argument for a divine being. Instead, a theology of nature uses the content of science to tutor, reformulate, and reinterpret traditional theological doctrines rather than to argue for the existence of God. Within the Christian intellectual community, for example, the doctrines of creation, providence, and human nature are affected in fascinating and important ways by the most influential scientific theories. As biochemist and theologian Arthur Peacocke states, for example, the traditional picture of nature as static and hierarchical has been replaced by a picture of nature as dynamic and developmental. For Peacocke, the new scientific picture presents an opportunity for an informed theology to assert that "the natural causal creative nexus of events is itself God's creative action."[24] Many see

---

[24] A. Peacocke, *Intimations of Reality* (Notre Dame, IN: University of Notre Dame Press, 1984), 63.

Peacocke's approach as opening the door for a closer relationship between religion and the most current science than seems possible in natural theology because it interprets the objects of science (entities and processes in the physical world, even characterized in a general way) as mediating God's presence and activity in the world.

A third version of integration is systematic synthesis, which tries to incorporate key theological themes and basic scientific insights into a total worldview. After all, the classical understanding of "uni-verse" is that all things are part of "one truth" – or, better, that all truths, in principle, fit together harmoniously, even if we do not always see how they fit in practice.

Nonetheless, a worldview provides a comprehensive framework that serves to fit all truths together in relationship. A well-known example of seeking a scientifically informed worldview in this way is the work of Alfred North Whitehead, who tried to create a total conceptuality that would harmonize religion, education, the arts, and human experience. Whitehead's "process philosophy" replaced the fixed, deterministic processes of Newtonian science with the concepts of change, randomness, and uncertainty characteristic of contemporary science – thus projecting the ideas that nature is open, relational, ecological, and interdependent. God, for Whitehead and his intellectual followers, is not the personal being of traditional religion but rather is a cosmic principle that seeks ideal aims for the universe and then optimally synthesizes the actual outcomes of all events in the history of the universe.[25]

Biology among the Sciences

After our preliminary survey of the nature of religion, the nature of science, and the kinds of philosophical questions that arise regarding their relationship, it is now appropriate to locate more precisely the scientific fields playing into the discussion that occupies this book. First, although we must inevitably speak of the sciences generally in making various broad points, our developing discussion centers on the science of biology specifically. It is typical to classify biology within the general discipline of science by

[25] A. N. Whitehead, *Process and Reality*, corrected ed. (New York: Free Press, 1985).

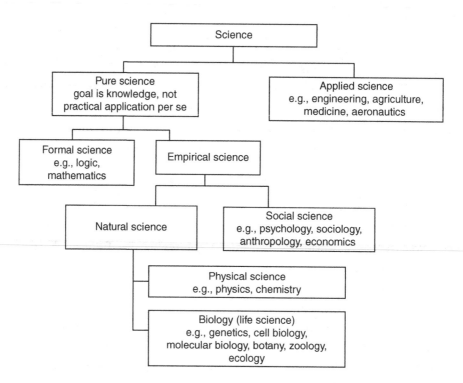

Figure 1.1 Areas of Science

envisioning it as a natural, empirical science. Since the term "biology" comes from two ancient Greek words – *bios* (life) and *logos* (words, organized thoughts, knowledge) – biology is a science of the living world. While there are other ways to classify the sciences, the present conception of biology can be defined and situated according to Figure 1.1.

On the one hand, all of the previously mentioned issues pertaining to the relationship between science and religion recur in our discussion of the relationship between biology specifically and religion. On the other hand, the field of biology raises some new and difficult issues related to religion that do not arise in regard to the sciences of the nonliving world.

There is no greater historical example of one who saw the interplay between biology and philosophy, and the implications for theology, than Charles Darwin himself, the great field biologist and scientific naturalist who thought very philosophically about his own discipline – acting like a philosopher of biology, as we would say today. Moreover, he characterizes his 1859 *Origin of Species* as "one long argument" for evolution by "natural selection."[26] He also offers in that monumental work a philosophical analysis of what counts as adequate biological explanation and a blueprint for the systematization of the biosphere. Darwin's contemporaries even referred to him as a philosopher – a philosopher, of course, who ended up providing the very foundation for biological science today. It is no wonder that an influential article by biologist Theodosius Dobzhansky was entitled "Nothing in Biology Makes Sense Except in the Light of Evolution."[27]

Although our classification of the science of biology is rather traditional, the biology within that taxonomy long ago became firmly anchored in evolutionary theory and findings, and thus developed into what we may call the "new biology," which has flourished in recent decades. Although Ludwig Boltzmann once insightfully remarked that the nineteenth century would be the century of Darwin in science, under Darwin's influence the biological sciences also experienced burgeoning growth in the second half of the twentieth century. How we perceive the living world and human life itself has thereby been greatly altered.[28]

In 1953, a major facet of the "modern synthesis" was revealed, a culmination of decades of combining neo-Darwinian biology with physics and chemistry. Work by James Watson and Francis Crick showing that DNA molecules have a three-dimensional chemical structure – suitable both for the faithful transmission of hereditary information and for the (heritable) variation necessary for Darwinian natural selection – was a major accomplishment in relating work by Mendel and Darwin.[29] Indeed, in the early 1900s, there had been concern that the newly rediscovered work of Mendel, with its emphasis on faithful transmission of hereditary information, might be incompatible with Darwinian natural selection, which required variation to arise. The discovery of the structure of DNA resulted from work that progressively showed the compatibility of Mendelism and Darwinism.[30]

---

[26] C. Darwin, *On the Origin of Species* (London: Murray, 1859), 459.

[27] T. Dobzhansky, "Nothing in Biology Makes Sense Except in the Light of Evolution," *The American Biology Teacher*, 35, no. 3 (1973), 125–129.

[28] L. Boltzmann, "The Second Law of Thermodynamics," in B. McGuinness (ed.), *Theoretical Physics and Philosophical Problems: Selected Writings* (Heidelberg: Springer, 1974), 15.

[29] J. D. Watson and F. H. C. Crick, "Molecular Structure of Nucleic Acids: A Structure for Deoxyribose Nucleic Acid," *Nature*, 171 (1953), 737–738.

[30] A. Stoltzfus and K. Caleb, "Mendelian-Mutationism: The Forgotten Evolutionary Synthesis," *Journal of the History of Biology*, 47 (2014), 501–546.

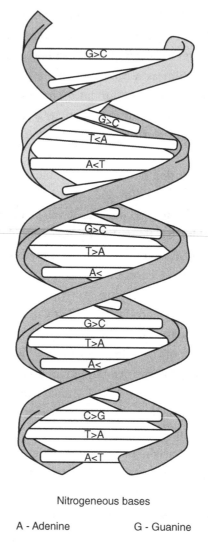

Nitrogeneous bases

A - Adenine   G - Guanine

T - Thymine   C - Cytosine

Figure 1.2 DNA Molecule

Now there is hardly a more famous symbol of the new biology than the famous double helix (Figure 1.2).

After the discovery of the structure of this amazing molecule, the resulting explosion of molecular biology was swift and powerful, transforming biology. When it comes to relating the new biology – its theories, findings, and potential implications – to religion, we are still engaging in a distinctively philosophical task that proceeds by raising fundamental questions, insisting on conceptual clarity, and rationally evaluating alternative views. Philosophy of biology, as a subdiscipline within the overall philosophy of science, is clearly involved, but philosophy is also interested in relationships between disciplines such as biology and theology. Metaphysics (which asks what reality is like and what sorts of things exist) and epistemology (which asks how we can know about the things that exist) are major philosophical interests in this interdisciplinary study as we sort out the kinds of realities and modes of knowledge of biology and theology. Other areas of philosophy, such as philosophy of mind and value theory, are also relevant to some issues. But we are now ready to embark on our journey and inquire more deeply into the complex relationship between biology and religion.

2 The Origin and Nature of Life

Life has long been thought to be a special quality that distinguishes entities that possess it from entities that are dead or inert. And life has always been fundamental to religious and spiritual traditions. Indeed, the very term "spiritual" is derived from Latin *spiritus* for breath – and other languages such as Greek (*pneuma*), Hebrew (*ruach*), and Sanskrit (*atman*) contain words variously translated as "spirit" or "breath" or "soul." However, life is also fundamental to the science of biology. Indeed, *bios* in classical Greek means "life" – physical life – and is the root of the term "biology." For centuries, it was commonly accepted, in religion and general society, that life was a special factor somehow distinct from but animating the physical organism. However, in the early years of modern biology, a purely physical basis for life was proposed – a position that was quite controversial, particularly because it had important, ostensibly adverse implications for religious and spiritual traditions.

In the present chapter, we survey some of the central discussions about the origin and nature of life itself that occur in biology as well as in religion. From the seventeenth century to the early twentieth century, an important debate took place between *vitalism*, which is the view that life is an intangible element distinct from the physical, and *mechanism*, the view that life is a purely physical process. Although the debate is no longer waged in exactly these terms, we track how certain instincts behind these two basic perspectives continue to play out today, paying particular attention to their scientific and religious implications. These days, debates over the origin and nature of life relate largely to the adequacy or inadequacy of the mechanical model. Origin-of-life discussions generally pertain to the possibility of chemical evolution, while discussions of the subsequent development and diversification of life typically either stay within the long-standing creation–evolution debates or attempt to move beyond them.

Vitalism in Early Biology

Aristotle, who was actually the first biologist in the Western tradition, identified the essence of life with "soul" (Greek: *psuché*) and assigned different kinds of souls to plants, animals, and humans. He advocated *hylomorphism* – the idea that each living thing is a unity of body and soul – such that the complete explanation of any biological structure or function must address both the specific matter and the soul by invoking two types of cause, mechanical and purposive. Mechanical causation pertains to the material aspect, whereas the particular kind of soul involved pertains to the intrinsic, intangible, purposive drive of a given biological entity. However, with the emergence of modern science and its mechanistic approach, life increasingly began to be viewed as having a material basis that was fully amenable to mechanical explanation, just like all other objects of scientific study. Obviously, both scientific and religious interests converge on the question of the nature of life, although in the debates in early biology, supporters of vitalism and mechanism did not segregate cleanly along the lines of religious and nonreligious affinities for a variety of complex reasons.

With roots in Aristotle, vitalism asserted that a "vital spark" or "inner force" was the essence of life. Early vitalists included Francis Glisson in the seventeenth century and Caspar Friedrich Wolff in the eighteenth century. In the nineteenth century, Johann Blumenbach actually explained the empirical process of regeneration – for example, in freshwater hydra – by reference to a "formative drive" in all living matter, although he theorized that this peculiar power may be "formed by the combination of the mechanical principle with that which is susceptible of modification."[1] Jöns Jakob Berzelius, who helped lay the foundation for modern chemistry, argued that a regulative force must be present within organic matter in order to maintain its various functions.[2]

---

[1] C. Birch and J. B. Cobb, *The Liberation of Life: From the Cell to the Community* (Cambridge: Cambridge University Press, 1985), 76–78.

[2] A. Ede, *The Rise and Decline of Colloid Science in North America, 1900–1935: The Neglected Dimension* (Burlington, VT: Ashgate, 2007).

For much of the nineteenth century, Johannes Peter Müller's *Handbook of Physiology* was the leading text in the field, providing detailed descriptions of circulatory, lymphatic, respiratory, digestive, and other systems in animals while supporting the vitalist contention that living organisms possessed a life energy or life force that could never be fully explained mechanically by reference to the laws of inorganic matter. Nonetheless, various scientific debates in the nineteenth century had major implications for the general idea of vitalism. The debate between supporters of "preformationism" and "epigenesis" is a case in point. Preformationism was a widely held theory that organisms develop from miniature versions of themselves instead of developing by the orderly buildup of component parts. Epigenesis, the denial of preformationism, is essentially the process by which animals, plants, and fungi develop from an egg, seed, or spore via a sequence of cell differentiation into body parts.[3] Embryologists eventually settled the debate against preformationism and in favor of epigenesis, which in turn made vitalism more difficult to defend.

Another related debate in the nineteenth century was waged over "spontaneous generation" and "biogenesis." At the time, spontaneous generation was commonly expressed (rather unreflectively) as the idea that maggots could spontaneously arise from dead flesh and fleas from dust, but more fully it reflected the general philosophical view that living things can arise from nonliving matter. Louis Pasteur, the father of microbiology and early proponent of germ theory, performed a famous experiment in which two flasks of nutrients were sterilized; then one flask was sealed and the other left open. The result was that microorganisms grew only in the open flask. Although Pasteur's experiment is sometimes mistakenly viewed as opposing vitalism, it rebuts only the "spontanist" version of vitalism. Because of Pasteur's work, the concept of biogenesis – that living things come only from living things – is traditionally attributed to him and has become an unquestioned hallmark of biology. Nevertheless, it is noteworthy that advocates of spontaneous generation, which was soundly refuted in the nineteenth century, operated on a fundamental instinct – that living things can somehow come from nonliving things. In the twentieth century, this same instinct was incorporated into the dominant mechanistic view of

---

[3] E. Mayr, *This Is Biology: The Science of the Living World* (Cambridge, MA: Harvard University Press, 1998), 11.

biology, which holds that life from nonlife is possible under the right conditions, an idea we discuss later in this chapter.

Interestingly, religious interests intersected with vitalism in different
ways. Pasteur was a Catholic Christian who held an independence view of
the relation of science and religion:

> In each one of us there are two men, the scientist and the man of faith or of
> doubt. These two spheres are separate, and woe to those who want to make
> them encroach upon one another in the present state of our knowledge![4]

Pasteur was an avowed freethinker who let the results of scientific experimentation and investigation speak for themselves – as in his famous experiment. At the turn of the twentieth century, as vitalism was falling out of
favor, a kind of neovitalism was championed by the German biologist and
embryologist Hans Driesch, who called the life force "entelechy," borrowing
Aristotle's term denoting a nonspatial "mindlike" force driving biological
life. Among other empirical evidences for entelechy, according to Driesch,
was the persistence of embryonic development despite interferences.
Zoologist Herbert Spencer Jennings, who was dedicated to the increasingly
accepted mechanical model in biology, denied that Driesch's proposal had
any explanatory value, arguing that entelechy "does not help our understanding of matters in the least."[5] Offering a middle way between vitalistic
and mechanistic thinking in biological science, French philosopher Henri
Bergson spoke of the *élan vital* – the "vital force" or "vital impulse" that is
immanent within all organisms and the creative impetus behind evolution.

The Mechanical Model in Modern Biology

Vitalist proposals receded into the minority as mechanistic thinking in
science increased, including mechanistic thinking in biology. Early in
seventeenth-century science, Descartes and his intellectual successors
extended mechanistic explanations of natural phenomena to biological
systems. Descartes himself advocated a substance dualism between mind
(*res cogitans*, thinking substance) and body (*res extensa*, extended substance),

---

[4] Pasteur in P. Debré, *Louis Pasteur*, trans. E. Forster (Baltimore: Johns Hopkins University
Press, 2000), 368.
[5] H. S. Jennings, "Behavior of the Starfish, Asterias forreri de Loriol," *University of California
Publications in Zoology*, 4 (1907), 180.

which led him to maintain that animals, including the human body, were "automata," mechanical devices differing only in degree of material complexity.

To be sure, vitalists during this time continued to point out problems with mechanistic thinking, arguing both scientifically and philosophically that Cartesian mechanism could not explain important biological features such as movement, perception, and the development of life. In experimental physiology, for example, the eighteenth-century anatomist Xavier Bichat studied types of living tissue and concluded that their "vital properties" were not purely physical properties and thus could not be totally explained by the mechanical model. Since the behavior of living tissue is irregular with respect to the mechanical forces recognized by Newton, Bichat reasoned that there must be other equally important forces of life – "sensibility" (responsiveness to the environment) and "contractility" (the ability of muscles to contract) – that were on par with Newton's forces. In fact, inorganic nature tended to tear down living organisms such that vital properties are required to oppose the effect of the inorganic. Bichat consequently characterized life as "the sum of all those forces which resist death."[6] He wrote, "To create the universe God endowed matter with gravity, elasticity, affinity, etc., and furthermore one portion received as its share sensibility and contractility."[7] Although Bichat argued that there were distinct properties of living tissues that could not be treated like objects of physics, the aim of mechanistic thinkers was to show how these properties of living systems arise out of the physical properties of the organism. For instance, François Magendie, in the mid-nineteenth century, argued that Bichat's living properties were actually "functions" of the underlying physical properties but were still not fully explainable in terms of the basic physical properties.[8]

Although many vitalists were accomplished experimentalists – most notably, Pasteur and Driesch – mechanistic criticisms accumulated and finally destroyed vitalism by the early twentieth century. In 1911, the biologist J. W. Jenkinson criticized vitalism by reminding the scientific

---

[6] X. Bichat, *Recherches Physiologiques sur la Vie et la Mort* (Paris: Marchant, 1805), 1. See W. Bechtel and R. C. Richardson, *Discovering Complexity: Decomposition and Localization as Strategies in Scientific Research* (Cambridge, MA: MIT Press, 2010), 102.

[7] X. Bichat, *Anatomie générale appliquée à la physiologie et à la médicine* (Paris: Brossom, Gabon et Cie., 1801), vol. I, xxxvii. See Bechtel and Richardson, *Discovering Complexity*, 102.

[8] F. Magendie, *Leçons sur les phénomènes chimiques et vivants* (Paris: Baillière, 1842).

community of the Ockhamist rule that their explanations should not postulate more entities than necessary to explain a given phenomenon.[9] Jenkinson's criticism was particularly effective because alternative mechanistic explanations for a number of biological structures and functions were being accepted. In 1948, philosopher of science Carl Hempel argued that, in postulating unobservable entities to explain life and its processes, vitalists thereby "render[ed] all statements about entelechies inaccessible to empirical test and thus devoid of empirical meaning."[10] In other words, no predictions could be inferred from the vitalist hypothesis and thus no confirmation or disconfirmation of it was possible.

Over time, the mechanical model became the new norm in biology. In modern biology, final causes – considered as reflecting the life force or pursuing goal-directed function – were replaced by mechanical causes, thus bringing biology into methodological unity with the rest of the natural sciences. Philosophically, Cartesian dualism, which sharply divides the realm of mind (which involves thought, agency, and purpose) from the realm of matter (which involves lawlike empirical regularities), became difficult to hold because increasingly successful mechanical explanations seemed to render purposive explanations unnecessary.[11] So biology gradually confined itself, like all other sciences, to the realm of matter operating by mechanistic causes rather than teleological causes. Even if dualism might be true, science going forward would deal only with matter and leave the nonmaterial realm to religion. In the words of E. J. Dijksterhuis, the great historian of the Scientific Revolution, God became a "retired engineer."[12] Of course, Enlightenment deism essentially held this exact position: that God created a world of laws and let them operate without interference. The world may once have been designed by God, but it was now to be seen as a machine, an ongoing clockwork system, run by unchanging laws. The picture of nature changed as the organic metaphor gave way to the machine metaphor.

---

[9] J. W. Jenkinson, "Vitalism," *The Hibbert Journal*, 9 (1911), 545–559.
[10] C. G. Hempel, "Studies in the Logic of Explanation," in C. G. Hempel (ed.), *Aspects of Scientific Explanation, and Other Essays in the Philosophy of Science* (New York: Free Press, 1965), 257.
[11] W. Bechtel and R. C. Richardson, "Vitalism," in E. Craig (ed.), *The Shorter Routledge Encyclopedia of Philosophy* (London: Routledge, 2005), 1051.
[12] E. J. Dijksterhuis, *The Mechanization of the World Picture* (New York: Oxford University Press, 1961).

Philosophical Issues in the Debate

A number of philosophical issues of particular interest to religion arise in the debate over vitalism and mechanism. One salient issue concerns the very nature of science – whether its method is restricted to the natural world and, if so, whether this means that science favors philosophical naturalism over any religious perspective. Another issue concerns whether biological properties can be reduced to material properties – which is essentially the debate between "reductionism" and "emergentism." Yet one more issue pertains to whether all relevant biological phenomena can be explained mechanically or if some phenomena might require teleological explanation.

Historically, the rejection of vitalism helped clarify the nature of science. Modern science operates according to "methodological naturalism" – that is, the procedure of seeking natural causes for natural phenomena, without appeal to any nonnatural or supernatural force or agency, and then codifying confirmed connections between natural causes and effects in terms of scientific laws. Methodological naturalism – by which science keeps pushing and pushing to make ever more discoveries – lies at the heart of the tremendous success of science. Yet methodological naturalism in science is controversial for some religious critics, who see it as inherently antireligious and atheistic because it eliminates God from nature and thus intentionality and purpose from our explanations of nature. The influential Christian philosopher Alvin Plantinga, for example, argues that methodological naturalism is a creature of Enlightenment rationalism that unfairly tilts the rational scales against religion and theism.[13] Evolutionary psychologists, for example, conclude that morality is an illusion perpetrated by our genes rather than something objective specifying how we should behave, as Christianity and many religions teach. Plantinga asks rhetorically, "Have they discovered, somehow, that Christian belief is in fact false?"[14] Thus, he argues that scientists should be able to consult their background beliefs – including religious beliefs – in their practice of science, including pursuing or preferring hypotheses that are compatible with Christian beliefs.

Defenders of methodological naturalism come in both religious and non-religious varieties. Many theists, for instance, see methodological naturalism

---

[13] A. Plantinga, *Where the Conflict Really Lies: Science, Religion, and Naturalism* (New York: Oxford University Press, 2011), 168–178.

[14] Plantinga, *Where the Conflict Really Lies*, 169.

as a divinely given procedure for gaining knowledge about the created natural world but do not insist that concepts of the supernatural have explanatory force within science. Francis Collins, the director of the successful Human Genome Project through the 1990s, is a scientist who employs methodological naturalism as a tool in his work and is also a Christian who believes that this mode of gaining knowledge was ultimately provided by God.[15] According to the Christian and theistic worldview that Collins embraces, methodological naturalism is the neutral procedure of science and can be utilized by any scientist, regardless of religious persuasion or lack thereof. The results generated are simply data for interpretation, in his case by a Christian worldview.

Most nontheists, in opposition, maintain that methodological naturalism strongly supports "philosophical naturalism" as a worldview, along with its fundamental assumptions that there is no God and that nature is the only reality. Philosopher of biology and Darwinian naturalist Michael Ruse argues that methodological naturalism is the correct procedure for acquiring knowledge of the physical world. But he presses beyond this epistemological point to argue the metaphysical point that methodological naturalism supports naturalism over theism, even if the version of theism is modest, like that of Francis Collins, and does not make frequent appeals to miracles to explain natural occurrences:

> [H]owever moderate your theism, methodological naturalism still inclines one to metaphysical naturalism. If you can explain the world without God, then as Pierre-Simon Laplace responded to Napoleon when asked why in one of his books he made no reference to the deity: "Sire, I had no need of that hypothesis."[16]

For many nontheistic thinkers involved in current science–religion debates, the combination of methodological naturalism and metaphysical naturalism forms a compelling intellectual package and indeed a total worldview. Biologist Richard Dawkins may first have coined the term "universal Darwinism" for his view that natural selection must be pervasive throughout the universe, but philosopher Daniel Dennett then extended its range to

[15] F. Collins, *The Language of God: A Scientist Presents Evidence for Belief* (New York: Free Press, 2006), 5.

[16] M. Ruse, "The Naturalist Challenge to Religion," in M. L. Peterson et al., *Philosophy of Religion: Selected Readings*, 5th edn. (New York: Oxford University Press, 2014), 433.

cover all of aspects of life – describing natural selection as acting like a "universal acid," unable to be contained, spreading out and transforming all domains of life, including morality and meaning.[17]

Typically, thinkers who combine methodological naturalism and metaphysical naturalism embrace some form of materialism: the idea that reality is nothing more than a collection of material entities in various configurations. Materialism typically leads to reductionism: the idea that the complex material entities that are studied by any science may in principle be reduced to their more basic material constituents, which are the purview of physics. Philosopher of science David Papineau endorses "the completeness of physics" – the principle that all phenomena may be broken down to the basic units of matter, which are currently quarks.[18] Biology in the middle of the twentieth century witnessed a migration of many researchers trained in physics to a "physics of biology" program. Francis Crick, the codiscoverer of the structure of the DNA molecule, declared that "the ultimate aim of the modern movement in biology is in fact to explain *all* biology in terms of physics and chemistry."[19] Physicist Erwin Schrödinger extended this thinking about DNA, leading to the "information" perspective on genes as bearers of genetic coding.[20] Famous biologists, such as J. B. S. Haldane and Conrad Waddington, welcomed the merger of physics and biology. Applied mathematician Norbert Wiener proposed that complex systems – from machines to cells – were fundamentally similar in being controlled and regulated by various feedback processes. Physicist George Gamow spoke of "information transfer in the living cell," and biologists François Jacob and Jacques Monod formulated an early theory of gene regulatory networks.

For perspective, we must distinguish "methodological reductionism," which is the legitimate scientific procedure of analyzing systems into their constituent units in order to understand them better, from "ontological reductionism," which is the philosophical position that nothing exists beyond the material composition of things. Crick and many other biologists had no religious or philosophical commitments that might deter them from associating their successful application of methodological reductionism in

[17] D. C. Dennett, *Darwin's Dangerous Idea: Evolution and the Meanings of Life* (New York: Simon & Schuster, 1995), 61–84.

[18] D. Papineau, *Philosophical Naturalism* (Oxford: Blackwell, 1993), chap. 1.

[19] F. Crick, *Of Molecules and Man* (Seattle: University of Washington Press, 1996), 10.

[20] E. Schrödinger, *What Is Life?* (1944; repr. Cambridge: Cambridge University Press, 2012).

their scientific work with a thoroughgoing ontological reductionism. The philosophical corollary of ontological reductionism is "epistemological reductionism" – the claim that the concepts employed in the various sciences are all reducible to the concepts of physics. When it comes to biology, then, the twin reductionist claims would be that biological entities are reducible to entities of physics and that biological concepts can be reduced to and understood in terms of the concepts of physics. Based on these reductionist views, many philosophers of science supported what is called "theory reduction," the view that the theories and laws in one field – in the present case, biology – are nothing but special cases of theories and laws of some more basic field in the physical sciences, and ultimately of physics.[21]

Ernst Mayr, the most eminent evolutionary biologist of the twentieth century, summarizes the procedure of theory reduction as follows:

> A theory T_2 (concerning a high level of organization) is reduced to a theory T_1 (concerning a lower level), if T_2 contains no primitive terms of its own, i.e., if the conceptual apparatus of T_1 is sufficient to express T_2.[22]

Theory reduction is widely discussed among philosophers of science, although scientists, on the whole, do not show great interest. Ernest Nagel provided the classical treatment of theory reduction, which is strongly supported by Kenneth Schaffner and Michael Ruse but rejected by David Hull and Philip Kitcher.[23] Ingo Brigandt and Alan Love have published a reasonably comprehensive and historically sensitive introduction to reduction for the biological sciences.[24]

However, despite the work of physicists on some biologically related problems, other biologically related issues have resisted reductionist efforts, such as problems pertaining to the genetic and biochemical mechanisms in molecular and cell biology. In fact, in 2004, Mayr published *What Makes*

[21] E. Nagel, *The Structure of Science* (New York: Harcourt, Brace & World, 1961); K. S. Schaffner, "Theories and Explanations in Biology," *Journal of the History of Biology*, 2 (1969), 19–33.

[22] E. Mayr, *What Makes Biology Unique?* (New York: Cambridge University Press, 2004), 78.

[23] Schaffner, "Theories and Explanations in Biology," 19–33; M. Ruse, "Reduction, Replacement, and Molecular Biology," *Dialectica*, 25, no. 1 (1971), 39–72; D. Hull, *The Philosophy of Biological Science* (Englewood, NJ: Prentice-Hall, 1974); P. Kitcher, "1953 and All That: A Tale of Two Sciences," *The Philosophical Review*, 93, no. 3 (1984), 335–373.

[24] I. Brigandt and A. Love, "Reductionism in Biology," in E. N. Zalta (ed.), *The Stanford Encyclopedia of Philosophy* (Stanford, CA: Spring, 2017), https://plato.stanford.edu/archives/spr2017/entries/reduction-biology/.

Biology Unique? – a major work arguing that biology is not reducible to physics but is rather its own special discipline. Mayr's rationale for the uniqueness of biology was that it is centrally concerned with function in organs and organisms, which are shaped by the exigencies of survival, thus raising the issue of goal-directedness. Hence, concepts of purpose or teleology, long banned from physics – because such concepts assume something nonempirical such as mind, intention, or agency – seem ineliminable from the modern biological sciences. Although the term "vitalism" has long since become anathema in biology, it is ironic that renewed talk of biological function and teleological explanation keep some of its concerns alive and present a serious obstacle to complete reductionism. Notably, in sciences such as evolutionary biology, genetics, medicine, and ethology, teleological concepts play an important explanatory role.[25] By contrast, advocates of mechanistic explanation argue that purpose is only apparent and can be reduced to mechanistic explanation, probably via concepts of "organization," an issue we discuss more fully in Chapter 3.[26]

The idea of organization is played out these days in what is known as "systems biology," where there are major critiques of essentially econometric, or abiotic, models of life that emphasize replicator dynamics over metabolic and homeostatic integration. At the beginning of the twenty-first century, the development of systems biology transformed the biological sciences: its new holistic approach treated organisms as integrated systems, thus supplying another major obstacle to reductionist approaches. The new approach transformed molecular biology into "systems molecular biology," which searches not for single molecules but for how molecular networks support biological structure and function. Interestingly, the rise of systems biology as a discipline actually raises the question of "emergence" – which is the appearance of new properties at the holistic or system level that are not

[25] See C. Allen and J. Neal "Teleological Notions in Biology," in E. N. Zalta (ed.), *The Stanford Encyclopedia of Philosophy* (Stanford, CA: Spring, 2020), https://plato.stanford.edu/archives/spr2020/entries/teleology-biology/; M. Artiga and M. Martínez, "The Organizational Account of Function Is an Etiological Account of Function," *Acta Biotheoretica*, 64, no. 2 (2016), 105–117; F. J. Ayala, "Teleological Explanations in Evolutionary Biology," *Philosophy of Science*, 37, no. 1 (1970), 1–15.

[26] A. Kolodkin, "Systems Biology through the Concept of Emergence," in S. Green (ed.), *Philosophy of Systems Biology* (Heidelberg: Springer, 2017), vol. XX, 181–191. See C. T. Wolfe, P. Huneman, and T. A. C. Reydon (eds.), *History, Philosophy, and Theory of Life Sciences*, 23 vols. (Dordrecht: Springer, 2013–2018).

(and cannot be) present in the parts.[27] J. B. S. Haldane, who did accept some form of teleology, remarked that "the doctrine of emergence ... is radically opposed to the spirit of science," but proponents assert that emergence is an empirical principle that pervades science irrespective of metaphysical commitments.[28] Thomas Henry Huxley's comment regarding the "aquosity" of water, a compound of hydrogen and oxygen, cites an example of emergence in the natural world. The particular properties of water so useful to biological systems are not easily predictable even when the properties of oxygen and hydrogen are reasonably well understood. Whereas reductionism might be able to give an account of the components – the most basic parts of the system – emergence considers the combination of factors necessary to explain the behavior of a dynamic system.

Philosophers of biology distinguish between strong emergence and weak emergence.[29] Weak emergence is the view that though predicting the features of a dynamic system from the properties of its constituent parts may be beyond current knowledge and ability, it is in principle possible to do so. Conversely, strong emergence states that genuinely new properties arise that cannot be predicted in advance and are not reducible to the properties of the constituent parts, nor to the interactions between them. Not surprisingly, the debate over strong versus weak emergence in biology centers on complex phenomena that are not fully understood. The transition from nonlife to life, the transition from single-celled life to multicellularity, and the transition from nonconsciousness to consciousness have all been proposed as examples of emergence, with some authors favoring strong emergence and others weak emergence.[30] Among working biologists, weak emergence is readily accepted as an observed phenomenon – there are many cases of dynamic systems that cannot, with current knowledge, be predicted from the ground up – but support for strong emergence is rare. This paucity of support for strong emergence may be due to the ongoing success of methodological naturalism and the pragmatic nature of biologists: since methodological

---

[27] Mayr, *What Makes Biology Unique?*, 74–77.

[28] J. B. Haldane, *The Causes of Evolution* (New York: Longmans, Green, 1932), 113.

[29] P. Clayton, "Conceptual Foundations of Emergence Theory," in P. Clayton and P. Davies (eds.), *The Re-emergence of Emergence: The Emergentist Hypothesis from Science to Religion* (New York: Oxford University Press, 2006), 7–8.

[30] See, for example, L. J. Rothschild, "The Role of Emergence in Biology," in Clayton and Davies, 162–164; and Chalmers, "Weak and Strong Emergence," in Clayton and Davies, 246.

naturalism continues to provide explanations for complex phenomena, there is no perceived need to reach for an alternative framework.

The Origin of Life

Although there is a technical distinction to be made between the prebiological and the biological, many biology textbooks include a section on the origin of life – the transition from nonlife to life – as a phenomenon falling broadly under the concept of evolution.[31] Furthermore, well-known evolutionists – such as Richard Dawkins, P. Z. Myers, and Nick Matzke – maintain that the origin of life is an aspect of evolution. Perhaps predictably, the debate within biology over whether life is the particular organization of an organism's constituent parts or is a unique quality somehow distinct from the material parts recurs in the study of life's origin. If certain chemicals (physical parts) can be arranged in a certain way, can they exhibit life and actually be alive? This is the assumption of the theorizing and research on "chemical evolution." Or is there something else distinct from the physical parts – some force or divine activity – that is required for a given piece of matter to be alive? Biology as a science traditionally assumes life (and that "life comes from life") and then proceeds to study its structures and functions, but abiogenesis research looks for the conditions that first produced life from nonliving matter. That transition point – that change from nonlife to life – presents an ultimate question for both science and religion.

To begin, let us consider the current state of abiogenesis science as well as its future prospects – in terms of theoretical proposals as well as any empirical work. As a scientific enterprise, work in abiogenesis must look for natural causes for natural phenomena. Although Darwin revolutionized biology by focusing on the living world, he also took as an assumption that natural, lawlike conditions were available for life's beginning. In a letter to botanist Joseph Hooker, he expressed the following thoughts on the origin of life:

> It is often said that all the conditions for the first production of a living
> organism are now present, which could ever have been present. – But if
> (& oh what a big if) we could conceive in some warm little pond with all sorts

---

[31] See G. A. Kerkut, *Implications of Evolution* (Oxford: Pergamon, 1960), 157.

of ammonia & phosphoric salts, – light, heat, electricity &c present, that a protein compound was chemically formed, ready to undergo still more complex changes, at the present day such matter w^d be instantly devoured, or absorbed, which would not have been the case before living creatures were formed.[32]

Darwin's theory of natural selection, which we discuss in the following chapter, was not relevant to the chemical precursors of life, since there is no competitive, selective environment until a replicating entity appears. However, many naturalists take Darwin's comment to embrace the possibility of a biochemical solution to the origin of life, which would clearly take its place under the mechanical model.

Modern mechanistic thinking about the origin-of-life problem dates from the 1920s when, independently, the Russian biochemist Alexander Oparin and the British biologist J. B. S. Haldane proposed an updated version of Darwin's hypothesis. They supposed that Earth's early atmosphere was reducing (i.e., limited in oxygen) such that organic molecules could form in that atmosphere (from ammonia and the action of ultraviolet radiation) and then, falling into the oceans, make for what Haldane called a "hot dilute soup" – that is, self-replicating molecule clusters could form under these conditions and life would be under way.[33] In the early 1950s, at the University of Chicago, when Stanley Miller and Harold Urey simulated the compound of chemicals in the early atmosphere and then administered regular electrical discharges, amino acids – the building blocks of proteins – were produced. Although this was not the creation of life per se, a major step seemed to have been taken in creating some essential molecules found in living beings from relatively simple precursors through plausible chemical mechanisms. This landmark experiment, though of only moderate scientific value today, inspired renewed interest for investigating the origins of life from a mechanistic perspective.

These mechanistic investigations have continued to bear fruit, though many unsolved questions remain. The origin of life is a frontier of science,[34] where several competing hypotheses have some support, and there is as yet

[32] Darwin to J. D. Hooker, February 1, 1871, Letter no. 7471, Darwin Correspondence Project, www.darwinproject.ac.uk/letter/DCP-LETT-7471.xml.
[33] J. B. S. Haldane, "Origin of Life," *Rationalist Annual*, 148 (1929), 3–10.
[34] D. R. Venema, "Intelligent Design, Abiogenesis, and Learning from History," *Perspectives on Science and Christian Faith*, 63, no. 3 (2011), 183–192.

no broad explanatory framework that makes sense of all observations. As science works to expand a theory from what is known into what is unknown, a frontier is expected. Given the success of explaining the diversification of life through the theory of evolution, scientists have hypothesized that similar mechanisms may have been at play at, or shortly after, life's origin. One key hypothesis is that life passed through an "RNA world" stage, either at its beginning or subsequent to it. Present-day life uses DNA for information storage and transmission, and proteins for most enzymatic activities, with RNA as an intermediary between them. RNA, however, can act as a hereditary molecule and as an enzyme, leading to the hypothesis that some form of RNA-based life preceded the current system – the so-called RNA world. Experimental support for this hypothesis has come in the last few decades from the increasingly successful production of RNA copiers by Jack Szostak's research team at Harvard University, among others.[35] The discovery that the ribosome, the enzyme that uses information copied from DNA to RNA to form proteins, uses only RNA for its enzymatic activity was widely seen as very strong support for this hypothesis.[36] In an important sense we still live in an RNA world – where RNA information and enzymatic activity is primary, and DNA and proteins can be seen as accessories to those primary functions. Recent work has also provided good evidence that the function of the ribosome – that of translating the genetic code from the language of nucleic acids to the language of proteins – is grounded in chemical attractions between amino acids (which make up proteins) and the groupings of nucleotides that code for them (so-called codons) in the current system. In other words, the hypothesis that the fundamental activities of present-day life may be reducible to chemical interactions has continued to find experimental support.[37]

Despite these tantalizing clues, however, many more mysteries remain. Of course, the production of various organic components and their organization is important to the mechanistic project, but solving the problem of life's origin also requires work at the level of a self-replicating entity and, later,

[35] R. F. Service, "A Newly Made RNA Strand Bolsters Ideas about How Life on Earth Began," *ScienceMag.org* (August 2016), www.sciencemag.org/news/2016/08/newly-made-rna-strand-bolsters-ideas-about-how-life-earth-began.

[36] T. R. Cech, "The Ribosome Is a Ribozyme," *Science*, 289, no. 5481 (2000), 878–879.

[37] M. Yarus, "The Genetic Code and RNA-Amino Acid Affinities," *Life*, 7, no. 13 (2017); D. B. F. Johnson and L. Wang, "Imprints of the Genetic Code in the Ribosome," *PNAS*, 107, no. 18 (2010), 8298–8303.

understanding the emergence of a fully functioning cell, the essential unit of present-day organisms. How metabolism arose (and even if it may have preceded nucleotide-based replication), how membranes came to be associated with life, and so on remain active areas of inquiry. Discussions continue about whether our models of the early atmosphere are correct and whether the early atmosphere was even the key to the start of life. With the development of plate tectonics, the hypothesis has become widely entertained that life started at the deep-sea vents, which spewed out chemicals at a high enough temperature to supply a source of energy. So, deep-sea vents – similar to Darwin's "warm little pond" – have been proposed as the location of original life. As a frontier of biology/chemistry, this diversity of hypotheses is to be expected and welcomed.

One proposal envisions a four-stage process for the chemical evolution of the first carbon-based life on Earth. Hypotheses for the right conditions that would have to be present include thermal vents, shallow surface ponds, sandy beaches, volcanic craters, clay deposits, and weathered feldspar. Under the right conditions, then, simple organic molecules concentrated and self-assembled into strings of nucleic and amino acids (RNA and proteins). Then, when a sufficient number of these molecules became concentrated together, they formed an interacting autocatalytic system that jointly catalyzed their mutual reproduction. Finally, these RNA-and-protein catalytic systems evolved, with RNA and eventually DNA taking on the role of information storage, which is what we observe today in all living cells. The process may be schematized as follows:

1. small organic molecules (e.g., amino acids, nucleic acid bases);
2. the small organic molecules combined to make larger biomolecules (e.g., proteins, RNA, lipids);
3. the larger biomolecules self-organized, by a variety of interactions, into a semi-alive system; and
4. the semi-alive system gradually transformed into a more sophisticated form, a living organism.[38]

Our task is not to follow the science in detail but to understand the project of finding a completely mechanistic explanation for an abiotic origin of life.

---

[38] L. Haarsma and T. M. Gray, "Complexity, Self-organization, and Design," in K. Miller (ed.), *Perspectives on an Evolving Creation* (Grand Rapids: Eerdmans, 2003), 228–310.

There have been advances as well as setbacks in finding a mechanistic explanation, with much remaining to be done. Each new hypothesis or discovery seems to multiply the problems, but no one would deny that that this is a fascinating and important field within the biosciences. A recent survey of this field by two leading researchers, Jeffrey Bada and Antonio Lazcano, is quite candid:

> Although there have been considerable advances in the understanding of chemical processes that may have taken place before the emergence of the first living entities, life's beginnings are still shrouded in mystery. Like vegetation in a mangrove swamp, the roots of universal phylogenetic trees are submerged in the muddy waters of the prebiotic broth, and how the transition from the non-living to the living took place is still unknown.[39]

Granting for the moment that this assessment of the state of abiogenesis research is fair, we can identify at least three broad reactions to origin-of-life research.

The first reaction, found mainly among naturalists, is one of optimism that a purely mechanical solution will be found – and thus that the origin of life would exhibit only weak emergence. As a philosophical naturalist, Michael Ruse, for example, holds that in principle there can be nothing but a mechanical solution because there are no causes but mechanical causes in the universe, which means that it is surely just a matter of time until scientific inquiry finds the mechanical solution.[40] It could, of course, be pointed out that Darwin's private notebooks indicate that the origin of life could be studied scientifically – for example, that "the intimate relation of Life with laws of chemical combination, & the universality of latter render spontaneous generation not improbable."[41] Although Darwin was not promoting folk vitalism, he was obviously recognizing that preexisting inorganic compounds were relevant to the emergence of life, although his central focus was on the development of life.

Other optimists about abiogenesis point to experimental success, such as the laboratory demonstration that RNA molecules can mutate, compete

[39] J. L. Bada and A. Lazcana, "The Origin of Life," in M. Ruse and J. Travis (eds.), *Evolution: The First Four Billion Years* (Cambridge, MA: Harvard University Press, 2009), 72.

[40] M. L. Peterson and M. Ruse, *Science, Evolution, and Religion: A Debate about Atheism and Theism* (New York: Oxford University Press, 2016), 85.

[41] Darwin Archive, Cambridge University Library, DAR122, 102e. Available online at http://darwin-online.org.uk/manuscripts.html.

against each other, and self-replicate, thereby setting up a selective situation that leads to evolution in an adaptive direction,[42] or the observed chemical affinities that seem to undergird ribosome function and origins.[43] Henderson James Cleaves of the Carnegie Institution for Science continues the optimism, saying, "Now making an artificial cell doesn't sound like science fiction any more. It's a reasonable pursuit."[44]

By contrast, a second naturalist position is pessimistic, maintaining that the nonlife/life boundary is absolute and cannot be crossed – and thus abiogenesis would be an example of strong emergence. In the mid-twentieth century, the philosopher of science Karl Popper expressed this view: "We may be faced with the possibility that the origin of life (like the origin of the universe) becomes an impenetrable barrier and a residue to all attempts to reduce biology to chemistry and physics."[45] More recently, atheist philosopher Thomas Nagel has denied the possibility that life could have emerged purely through the actions of physics and chemistry: "My skepticism is not based on religious belief, or on a belief in any definite alternative. It is just a belief that the available scientific evidence, in spite of the consensus of scientific opinion, does not in this matter rationally require us to subordinate the incredulity of common sense. That is especially true with regard to the origin of life."[46]

A third reaction proposes, at least in a general way, that there is some other factor in the origin of life besides the natural. Nagel, for instance, who is an atheist, goes on to argue that nature is infused with life – consciousness being just the most visible end of the spectrum – and so the problem becomes a kind of nonproblem. Interestingly, his claim that this problem is impossible for the mechanical model and thus invites some kind of teleology makes one wonder how he or anyone can remain a thoroughgoing atheist if life is not thought to have a purely mechanical solution. Perhaps even more interesting is the fact that Antony Flew, a dominant atheistic

[42] E. Eckland, J. Szostak and D. Bartel, "Structurally Complex and Highly Active RNA Ligases Derived from Random RNA Sequences," *Science*, 269 (1995), 364–370.
[43] Johnson and Wang, "Imprints of the Genetic Code in the Ribosome."
[44] H. J. Cleaves in C. Zimmer, "On the Origin of Life on Earth," *Science*, 323, no. 5911 (2009), 198–199.
[45] K. R. Popper, "Scientific Reductionism and the Essential Incompleteness of All Science," in F. J. Ayala and T. Dobzhansky (eds.), *Studies in the Philosophy of Biology: Reduction and Related Problems* (London: Macmillan Education, 1974), 270.
[46] T. Nagel, *Mind and Cosmos: Why the Materialist Neo-Darwinian Concept of Nature Is Almost Certainly False* (New York: Oxford University Press, 2012), 7.

voice in philosophy in the second half of the twentieth century, converted to a position like deism near the end of his life because he developed doubts about a thoroughly mechanical solution: "The present physicists' view of the age of the universe gives too little time for these theories of abiogenesis to get the job done. ... The [unanswered philosophical question is, how can] a universe of mindless matter produce *beings with intrinsic ends, self-replication capabilities, and 'coded chemistry'?*"[47]

Religious thinkers who share this third type of reaction affirm that the origin of life is due to divine activity, although they differ over whether that activity was direct or indirect. According to some religious approaches, God acted directly, miraculously, to give life to physical organisms; for other religious approaches, God acted through natural processes that are being discovered by science. Interestingly, all three positions on the prospects for a scientific answer to the origin-of-life question also apply to the broader question of the development of life, to which we now turn.

The Creation–Evolution Controversy

A wide variety of biological phenomena – from the diversification of species to the organization of the cell – require explanation. Is an evolutionary – and thus mechanical – explanation sufficient? Or must there be some sort of divine involvement? At the deepest level, answering these sorts of questions calls upon our fundamental worldview assumptions. Consider how the following passages from Genesis in the Judeo-Christian scriptures provide a touchstone for revealing differences among worldview approaches to biology:

> And God said, Let the earth bring forth grass, the herb yielding seed, and the fruit tree yielding fruit after his kind, whose seed is in itself, upon the earth: and it was so. (Gen. 1:11)

> And God made the beast of the earth after its kind, and cattle after their kind, and every thing that creepeth upon the earth after his kind: and God saw that it was good. (Gen. 1:25)

---

[47] A. Flew, *There Is a God: How the World's Most Notorious Atheist Changed His Mind* (New York: HarperCollins, 2007), 123.

So God created man in his own image, in the image of God created he him; male and female created he them. (Gen. 1:27)

And the Lord God formed man of the dust of the ground, and breathed into his nostrils the breath of life; and man became a living soul. (Gen. 2:7)

Christian groups that accept these verses literally have a more direct view of God's role in the origin of living things – for example, the phrase "after its kind" is taken to mean "fixity of species, with no evolution" and the "creation of male and female" to mean without an animal prehistory. Other Christian traditions take these verses as elements in a symbolic narrative that conveys the theological theme that God is the ultimate source of all creation and all life. Thus, they believe that God was involved but assert that his will was accomplished indirectly through natural processes. In Islam, another monotheistic religion, we again find a range of opinions, from the more literalistic to the more symbolic. For instance, the Qur'an indicates that "Allah has created from water every living thing,"[48] a statement subject to debate among Muslim scholars.

According to historian of science Ronald Numbers, we cannot underestimate the influence of the literalist interpretation of Genesis on the cultural discussion of science – particularly evolutionary science – in American culture. His book *The Creationists* explains that "creationism" is rooted in early twentieth-century Protestant fundamentalism, which insists that the Bible is "inerrant" and therefore must be interpreted literally. The typical creationist position – called "Young Earth Creationism" – asserts a six-day creation, a young Earth 6,000 to 10,000 years old, the instantaneous creation of all animal "kinds" and their subsequent rapid diversification to present forms, the special creation of the first humans Adam and Eve, and a global flood. Furthermore, the early environment of life was pristine and perfect, and without sin or pain or death, a subject we will explore more fully in Chapter 4.

Protestant fundamentalism is clearly in conflict with modern science because it rejects many established scientific facts. Nonetheless, what became known as "Creation Science" or "scientific creationism" developed as an arm of biblical creationism – as a kind of alternative science that corrects mainline science and construes certain aspects of science to favor

---

[48] The Qur'an 21:30, see also 24:45.

its position. The Creation Research Society was formed in 1963 and led to the founding of a center for creation research in 1970, which was eventually reorganized into the Institute for Creation Research (ICR) by Henry Morris, a hydraulic engineer.[49] However, the most widely known Creation Science operation these days is the rather high-tech Creation Museum near Cincinnati, Ohio, founded in 2007 by Ken Ham, an apologist for biblical literalism.[50] An instructive and well-known example of a Creation Science argument was the insistence that there are "missing links" in the fossil record, making evidence for evolutionary common ancestry (particularly leading to humans) sadly inconclusive. However, from Darwin on, evidence for common ancestry steadily mounted in fields such as comparative morphology, archaeology, and paleontology, only to be greatly augmented and confirmed in recent decades by DNA comparisons in contemporary molecular genetics. For example, since the completion of the Human Genome Project in 2003, we have compared the human genome to chimpanzee and gorilla genomes and have found that all three genomes have exactly the features one would expect if they share a common ancestor. All three genomes are over 90 percent identical, with the same genes in the same order, and with numerous features that are indicative of shared ancestry.[51]

Both secular and religious critics have consistently made the philosophical point that Creation Science is not bona fide science – that it is instead pseudoscience – because its methods generate empirically untestable hypotheses about supernatural causes. Nonetheless, in the past 100 years or so, as Creation Science developed into a cultural and political force, the correct definition of what counts as science has become a critical issue. Technically, this is known as the "demarcation problem" of distinguishing science from nonscience. Because fundamentalists through the years have objected to teaching evolution in public schools and demanded at least equal treatment of Creation Science or its equivalent, a number of court cases have been generated that revolve around the true nature of science. In 1925,

---

[49] R. Numbers, *The Creationists: From Scientific Creationism to Intelligent Design*, expanded ed. (Cambridge, MA: Harvard University Press, 2006), 220, 229, 312–315.

[50] S. L. Trollinger and W. V. Trollinger, *Righting America at the Creation Museum* (Baltimore: Johns Hopkins University Press, 2016), chap. 1.

[51] D. R. Venema, "Genesis and the Genome: Genomics Evidence for Human-Ape Common Ancestry and Ancestral Hominid Population Sizes," *Perspectives on Science and Christian Faith*, 62, no. 3 (2010), 166–178; D. R. Venema and S. McKnight, *Adam and the Genome: Reading Scripture after Genetic Science* (Grand Rapids MI: Brazos Press, 2017).

the famous Scopes Monkey Trial in Dayton, Tennessee, found biology teacher John Scopes guilty of teaching evolution in violation of state law – a ruling later overturned by the Tennessee Supreme Court on grounds that evolution is science, whereas creationism is not.[52] Although other court cases have also involved the idea of special creation in public school biology classes, one case in a US federal court appears to have brought legal closure to the issue. In 2004, when the Dover School District in Pennsylvania required that "intelligent design" be taught as an alternative to evolution, plaintiffs argued in 2005 during *Kitzmiller* v. *Dover Area School District* that it was another form of religious creationism and violated the Establishment Clause separation of church and state. Judge John Jones ruled in favor of the plaintiffs, declaring that the school policy was unconstitutional because intelligent design is not science.[53] (Intelligent design will be treated more fully in Chapter 3.)

Although fundamentalist Christian groups envision God's direct miraculous activity in the appearance of life, other Christian perspectives embrace a range of metaphysical theories regarding the indirect activity of God in creation. Concepts of divine energy, divine impartation of a supervenient quality, and a divine method of working through secondary causes are all possible ways of thinking about how an infinite, living, conscious being is the source and sustainer of finite living beings. While it is not within the purview of science to detect divine creative involvement, many believe that a religious metaphysical framework is required to make best philosophical sense of the living world that science investigates.[54]

The classic Thomistic tradition contains the best-known Christian affirmation of God's indirect activity in the world via natural causes. According to Aquinas, God as Primary Cause gives creation its various powers and capabilities and thus enables it to operate through "secondary causes." In essence, God's providence consists in the donation of created powers and the ordering of all created powers toward their proper ends. However, Aquinas carefully avoids total determination of all things by God; he argues that God's plan is for created beings to play a genuinely causal role in the unfolding development of the world. Since this view sees events in the natural world as

---

[52] E. Larson, *Summer for the Gods: The Scopes Trial and America's Continuing Debate Over Science and Religion* (New York: Basic Books, 2008).

[53] *Kitzmiller* v. *Dover Area School District*, 400 F. Supp. 2d 707 (M.D. Pa. 2005). See Numbers, *The Creationists*, 391–394.

[54] See Peterson and Ruse, *Science, Evolution, and Religion*, 49–51.

divinely given potentialities becoming actual, it does not entail that theology is either in conflict with or independent of science. Instead, Aquinas's perspective is meant to provide a metaphysical and theological framework for science and the world it investigates.[55]

The Thomistic view that created potentialities can activate, actualize, and evolve is relevant to all physical phenomena, and this view is particularly relevant to biological life and its goal-oriented behavior, which are physical phenomena that appear qualitatively distinct from nonliving physical phenomena. For Aquinas, a living thing is "animated" by a "soul" (Latin: *anima*; Greek: *psyche*). Regarding the nature of the soul, Aquinas states the following:

> [T]he soul is said to be the first principle of life in the things that are alive around us. For we say that living things are animate, whereas inanimate things are those without life . . . . [T]he ancient philosophers [mistakenly] claimed that the principle behind these functions is a body. They said that the only things that exist are bodies, [that] the soul is a body.[56]

On this view, every type of living thing – whether vegetable, animal, or rational – has its own unique immaterial soul, its "first principle" or "moving principle," that is distinct from its material composition.

Darwin in the *Origin* actually indicates (much as Aquinas would) that it is better for the Creator to produce species by secondary causes that have their own powers to develop rather than by instantaneous special creation. Darwin wrote:

> Authors of the highest eminence seem to be fully satisfied with the view that each species has been independently created. To my mind it accords better with what we know of the laws impressed on matter by the Creator, that the production and extinction of the past and present inhabitants of the world should have been due to secondary causes, like those determining the birth and death of the individual. When I view all beings not as special creations, but as the lineal descendants of some few beings which lived long before the first bed of the Cambrian system was deposited, they seem to me to become ennobled.[57]

[55] Cf. I. G. Barbour, *Religion and Science: Historical and Contemporary Issues* (New York: HarperOne, 1997), 98–105, 309–312.
[56] Aquinas, *Summa Theologiæ*, 1.75.1.
[57] C. Darwin, *On the Origin of Species*, 2nd edn. (London: John Murray, 1860), 489.

No doubt, Darwin was thinking of contemporary naturalists like Richard Owen and Sir Charles Lyell, who thought that evolution could not be reconciled with their Christian beliefs.

Reflecting on the Mystery of Life

We commonly speak of "the mystery of life," but what kind of mystery is involved? Could there be different kinds of mystery? For science, a mystery is the challenge of developing an empirical explanation for some important but currently unexplained phenomenon. For religion in this context, a mystery involves a level of reality and meaning lying beyond science. Religion and science typically interface when science seems to reach some limit at which we can form a boundary question, as discussed in Chapter 1. The Big Bang in science, for example, raises boundary questions, such as Leibniz suggested: Why is there something rather than nothing? Why was there any matter and energy at all? Similarly, the biological phenomena associated with the origin and nature of life raise a number of limit questions: What is life? How did it begin? Why is there an amazing abundance of life-forms? Religion has an important stake in contributing answers to these questions, but naturalist-oriented viewpoints have an equal stake in countering religious explanations.

Some parties to the discussion trade on a dichotomy between mechanical and teleological explanations and between secular and religious visions. On the one hand, naturalist-oriented thinkers maintain the sufficiency of science to generate a mechanical explanation telling us what life is and how it began. Their confidence is that, given enough time, creativity, and funding, mechanical explanations of all biological phenomena will be found. On the other hand, fundamentalist creationist groups argue that a purely mechanical explanation is not possible because life is a special quality that requires a teleological explanation referring to the creative activity of God. Of particular interest for these groups is the status of humanity as a unique divine creation with a special relation to God (see Chapter 6).

As we saw earlier, the possibility of a middle way between these extreme views would recognize the success of science and yet hold that science requires philosophical and perhaps theological contextualization. In the present case, then, evolutionary biology should investigate as far as it can investigate, but its methods and findings will still require worldview interpretation. This perspective, rooted in Aquinas, has long been held by the

Catholic Church. In 2014, Pope Francis followed a succession of popes who have jointly affirmed evolutionary science and Christian faith: "Evolution in nature is not inconsistent with the notion of creation, because evolution requires the creation of beings that evolve."[58]

It seems that two opposing themes emerge regarding the present topic: either that the natural universe is capable of producing life or that it is not capable. When pondering the deep question of life itself, our direction for answering will be grounded in our worldview and influence how we evaluate all the other related questions. Some naturalists and materialists have been extremely certain of an affirmative answer, claiming that the natural universe can produce life. For instance, Christian Duve, a Nobel Prize winner in physiology, declared that "life is a cosmic imperative," essentially saying that nonlife can readily give rise to life.[59] However, those same naturalists and materialists hold that both the universe and life itself occurred by chance. The French biologist Jacques Monod maintained that evolution proceeds by "pure chance, absolutely free but blind."[60] Paradoxically, chance is generally understood to entail that life is staggeringly improbable but that in reality the improbable happened to occur. Christian believers, on the other hand, assert in various ways that God is behind life as we know it, that "God breathed" life into creatures. For these believers, biological life comes ultimately from a self-living being, God, who is addressed many times in the Anglican Book of Common Prayer as "the Lord, the Giver of Life."[61]

In a context of differing opinions, the ever-diplomatic Darwin stated that "there is grandeur in this view of life" that recognizes a Creator of the simplest of organic forms but affirms that the unfolding of organic life on this planet, including higher forms of life, occurred according to natural laws. He wrote, "[F]rom so simple a beginning endless forms most beautiful and most wonderful have been, and are being, evolved."[62] In the following chapter, we consider the order and organization of life, which raises another set of important philosophical questions.

[58] Francis, "Address of his Holiness Pope Francis on the Occasion of the Inauguration of the Bust in Honor of Pope Benedict XVI," October 27, 2014, www.vatican.va.

[59] C. de Duve, *Vital Dust: Life as a Cosmic Imperative* (New York: Basic Books, 1995), 300.

[60] J. Monod, *Chance and Necessity: An Essay on the Natural Philosophy of Modern Biology*, trans. A. Wainhouse (New York: Knopf, 1971), 112–113.

[61] *The Book of Common Prayer* (1789; repr. New York: Church Publishing Incorporated, 2007).

[62] Darwin, *Origin of Species*, 2nd ed., 490.

3   The Question of Design in Living Systems

Perhaps no issue at the intersection of biology and religion has received more attention now or in the past than the issue of design. The general problem is whether to explain the manifest appearance of design in nature as apparent or real. In living systems, should function be attributed to an intelligent cause or to natural causation? The debate over this question is as ancient as it is contemporary. Of particular historical and intellectual importance is the tension between the design argument advanced by many thinkers, famously presented by William Paley in the early nineteenth century, and the natural biological explanation, offered by Charles Darwin in the mid-nineteenth century. As our discussion develops, we evaluate the Intelligent Design (ID) argument of the past few decades as well as various forms of theistic evolution as ways of combining ideas of intelligence or purpose with the facts of biology. Likewise, we discuss and evaluate both scientific and philosophical perspectives that reject the idea that intelligence or purpose can be applied to biology or any of the other sciences.

Classical Discussions of Design

Discussions of biological teleology have very ancient roots. In the *Timaeus,* Plato depicts the divine Craftsman or Demiurge who fashions the world of matter according to eternal forms, transcendent Ideas of what everything, including every living biological organism, should be like. All things, including living beings, are directed toward the external good of the transcendent Ideas by a rational, purposeful, and beneficent agency. By contrast, Aristotle's teleological biology emphasizes the immanence of *telos*, or purpose, within the organism, a goal-directedness, we might say, that provides an intrinsic principle of change and development within the organism itself. Aristotle retained the notion of formal cause, which reflected Plato's approach, but included it as just one of four causes that made for complete explanation of any phenomenon, along with material, efficient, and final causes. In Aristotle's *Physics*, the final cause – the intrinsic *telos* or goal – directs the behavior of living beings.[1] While not a cosmic teleology like Plato's, Aristotle's teleology of organismal development, for example, was an impetus or goal-directed process acting as a principle of change within the organism, an inherent property driving its development. Plato's and Aristotle's views were teleological and very influential for many centuries on discussions in biology, but they were essentialist, holding that species are kinds that do not admit of anything we would call evolutionary change.

In the second century AD, Greek surgeon and philosopher Galen provided a good example of applying Aristotelian teleology in physiology, arguing that the "existence, structure, and functional analysis of all the parts must be explained by reference to their functions in promoting the activities of the whole organism."[2] For Galen, whose writings dominated medical thought until the seventeenth century, Aristotelian final causes provided an explanation for the parts of an organism that was superior to the explanation by mechanical causes. Teleological thinking, still following Aristotle, was deeply incorporated into both medieval Islamic and Christian religious traditions as an element of natural theology. As historian of ideas Arthur Lovejoy has pointed out, teleological thinking pervaded intellectual work in the medieval period. Everything in the perfectly ordered "Great Chain of Being" had its ordained place and pursued its essential purpose.[3] In the twelfth century, the great Islamic philosopher Averroës (Ibn Rushd) proposed a version of the teleological argument.[4] In the thirteenth century, Latin philosopher and Christian theologian Thomas Aquinas became the fountainhead architect of Christian natural theology, presenting, among other arguments, a broad version of the teleological argument – based on

---

[1] Aristotle, *Physics*, bk. 2.

[2] M. Schiefsky, "Galen's Teleology and Functional Explanation," in David Sedley (ed.), *Oxford Studies in Ancient Philosophy*, vol. 33 (2007), 371.

[3] A. O. Lovejoy, *The Great Chain of Being: A Study of the History of Ideas* (Cambridge, MA: Harvard University Press, 1936).

[4] Averroës, *Faith and Reason in Islam: Averroës' Exposition of Religious Arguments*, I. Y. Najjar (ed.) (Oxford: Oneworld, 2001), 33–38, 78–92.

the goal-directed activity observed in nature – as the fifth of his "five ways" of proving the existence of God.[5]

However, a noticeable pivot toward mechanistic thinking occurred in 1628 when the anatomist and physiologist William Harvey published *On the Motion of the Heart and Blood*.[6] Opposed to Galenic or Aristotelian approaches to anatomy, Harvey presented a mechanistic account for the structure and function of the heart without reference to final causes.[7] Of course, during this period Descartes provided a general philosophical rationale for applying a mechanistic approach to all of material nature. But seventeenth-century voices for teleology in biological nature were not totally silenced, as evidenced in the works of John Ray, a leading naturalist and botanist who made important contributions to taxonomy. In addition to his classification work on plants, he published *The Wisdom of God Manifested in the Works of the Creation*, which featured an argument from design based on the form and function of organic nature.[8]

However, by the end of the eighteenth century, Immanuel Kant had famously advanced the view that the human mind inevitably imposes teleology on living things but that we cannot know whether there is actually teleology in the natural world. Yet, as Kant admitted, organisms exhibit capacities to grow and reproduce that trigger teleological interpretation while creating puzzlement for the mechanical interpretation of science. Kant himself thought that "there will never ever be a Newton of the blade of grass" – a comment widely taken to mean he thought it impossible to bring mechanistic explanation to the living world studied in biology.[9] Nonetheless, the momentum toward exclusively mechanistic explanations increased, as shown by the nineteenth-century attempt within organic chemistry to determine whether living things were nothing more than complex chemical systems.[10]

---

[5] Aquinas, *Summa Theologiæ*, 1.2.2.
[6] W. Harvey, *On the Motion of the Heart and Blood in Animals*, trans. R. Willis (1628, repr. New York: P. F. Collier & Son, 1910).
[7] R. K. French, *William Harvey's Natural Philosophy* (New York: Cambridge University Press, 1994).
[8] John Ray, *The Wisdom of God Manifested in the Works of the Creation* (London, 1691).
[9] I. Kant, *The Critique of Judgment* (New York: Hafner Publishing Company, 1951), 270.
[10] T. Lenoir, *The Strategy of Life: Teleology and Mechanics in Nineteenth-Century German Biology* (Chicago: University of Chicago Press, 1982).

Paley and Darwin

David Hume, the eloquent eighteenth-century skeptic, particularly criticized the prevalent design argument because it was technically a bad argument in general and its attempted analogy admits of telling disanalogies. Exposing the anthropomorphic and selective reasoning of the argument that purports to prove a single grand designer, Hume pointed out that "several deities" may have been involved because humans often cooperate in building projects and that the original designer might not even remain in existence after creating the world long ago.[11] But perhaps his most serious objection was based on the "faulty and imperfect" aspects of the world, which could argue for an infant, senile, or incompetent deity, in any case.[12] Of course, among other things, the "imperfections" include all the evil and suffering in the world, which means that the problem of evil (covered in Chapter 4) is a major rational obstacle to the argument from design in natural theology. But regardless of its inherent philosophical difficulties, design-type thinking in biological science persisted since there was no alternative scientific explanation for complex organic structures.

In the nineteenth century, the clash between teleological and mechanical explanations was epitomized in the approaches of two towering intellectual figures – natural theologian Archdeacon William Paley and scientist Charles Darwin. In his 1802 work of Christian apologetics entitled *Natural Theology or Evidences of the Existence and Attributes of the Deity*, Paley made use of the well-known and often-repeated watchmaker analogy for the existence of God, using the model of Newtonian scientific methodology. Paley constructs his classical argument for design from data of the natural world:

> In crossing a heath, suppose I pitched my foot against a *stone*, and were asked how the stone came to be there, I might possibly answer, that for any thing I knew to the contrary it had lain there for ever; nor would it, perhaps, be very easy to show the absurdity of this answer. But suppose I had found a *watch* upon the ground, and it should be inquired how the watch happened to be in that place, I should hardly think of the answer which I had before given, that for any thing I knew the watch might have always been there. Yet why should not this answer serve for the watch as well as for the stone; why is it not admissible in the second case as in the first? For this reason, and for no

---

[11] D. Hume, *Dialogues Concerning Natural Religion* (1779), 5.8.   [12] Hume, *Dialogues*, 5.12.

other, namely, that when we come to inspect the watch, we perceive – what we could not discover in the stone – that its several parts are framed and put together for a purpose, *e.g.* that they are so formed and adjusted as to produce motion, and that motion so regulated as to point out the hour of the day; that if the different parts had been differently shaped from what they are, or placed after any other manner or in any other order than that in which they are placed, either no motion at all would have been carried on in the machine, or none which would have answered the use that is now served by it.[13]

The watch analogy leads to "the inference we think is inevitable, that the watch must have had a maker."[14] This maker formed the watch for a purpose and designed its use. Paley's work proceeded to develop the argument by citing the complex organization of living things that further support the conclusion of an intelligent designer. For example, he argued that biological structures – such as the eye, which is "made for vision," with parts assembled to refract light rays and enable sight – are purposive mechanisms.

Paley's general aim was to prove God's existence from what God has already accomplished in ordering nature and then to argue that God's continuing benevolence was the best explanation of phenomena that science could not explain. Even Newton used God as the explanation for lacunae in his own mechanistic explanations – say, of the motion of Mercury, which was not precisely explained until the advent of relativity – but the "God-of-the-gaps" approach (a kind of teleology by default) declined in popularity and usefulness as science discovered more mechanistic explanations. A salient point of Paley's was that the "sovereign order of nature" – conceived according to the English Newtonian tradition – was opposed to the "blind chance" advocated by such thinkers as Erasmus Darwin, Charles Darwin's grandfather. Paley was representative of many religious thinkers of the day in seeing chance as linked philosophically to atheism and materialism and linked politically to anarchy as personified by the French revolutionaries.

Yet even before Darwin's discovery, both religious and nonreligious thinkers in the nineteenth century were coming to believe the evidence for species change and evolution. His special contribution was to identify a primary mechanism for evolution – natural selection – and thus to provide

---

[13] W. Paley, *Natural Theology: Or, Evidences of the Existence and Attributes of the Deity*, 12th edn. (1802, repr. London: Faulder, 1809), 1.
[14] Paley, *Natural Theology*, 3.

a plausible explanation for "transmutation of species," which had already been proposed by others, including the anonymously published *Vestiges of the Natural History of Creation*, which scandalized Victorian England in the decades prior to Darwin.[15]

In 1825, the Earle of Bridgewater, the Reverend Francis Henry Egerton, sponsored a series of eight books aimed at keeping natural theology alive by supporting teleological thinking in relation to new ideas in science, such as the transmutation of species. One of these treatises, *Astronomy and General Physics Considered with Reference to Natural Theology*, was written by mathematician, scientist, priest, and philosopher William Whewell, one of Charles Darwin's professors at Cambridge. Whewell was quoted by Darwin in the frontal material to the *Origin of Species*:

> But with regard to the material world, we can at least go so far as this – we can perceive that events are brought about not by insulated interpositions of Divine power, exerted in each particular case, but by the establishment of general laws.[16]

Science qua science must assume unbroken laws – which Whewell affirmed were established by God – and cannot countenance special miracles.

So it was actually Darwin who completed the destruction of Paley's argument as a viable piece of natural theology by finding an alternative, and purely natural, explanation to what was already seen as logically deficient reasoning. Darwin's discovery of natural selection as the mechanism of species adaptation and change brought biology in line with the mechanistic paradigm common in the other sciences. Darwin makes the point in his autobiography:

> The old argument from design in nature, as given by Paley, which formerly seemed to me so conclusive, falls, now that the law of natural selection has been discovered. We can no longer argue that, for instance, the beautiful hinge of a bivalve shell must have been made by an intelligent being, like the hinge of a door by a man. There seems to be no more design in the variability of organic beings, and in the action of natural selection, than in the course which the wind blows. Everything in nature is the result of fixed laws. But

---

[15] See, for example, P. Sloan, "Evolutionary Thought before Darwin," in E. N. Zalta (ed.), *The Stanford Encyclopedia of Philosophy* (Winter 2019), https://plato.stanford.edu/archives/win2019/entries/evolution-before-darwin/.

[16] W. Whewell, in C. Darwin, *On the Origin of Species* (London: Murray, 1859), ii.

I have discussed this subject at the end of my book on the *Variation of Domesticated Animals and Plants*; and the argument there given has never, as far as I can see, been answered.[17]

Although Darwin knew the design argument well, having been immersed in Paley's texts as a student at Cambridge and observing its widespread cultural influence, he felt constrained to report his scientific discovery and explain its impact on the key argument of English natural theology.

With natural selection as his "theory with which to work," Darwin explained a wide array of known facts – including adaptation, extinction, geographical distribution, and vestigial structures – that were problematic for all other theories, including those asserting specific divine design. It is noteworthy that chance, the very thing Paley and many others feared, played a central role in Darwin's theory of evolution through natural selection, as we shall see. First, Darwin argued for a "struggle for existence." Organisms have a natural tendency to reproduce at an explosive rate – following Malthus, Darwin thought that this rate would be geometric (1, 2, 4, 8, ...). However, since there is limited food and space, at most, as Malthus argued, resources can only increase at an arithmetic rate (1, 2, 3, 4, ...). Therefore, not all of the living creatures that are born can survive and, what is even more important in the biological world, not all can reproduce. There is going to be a struggle, although, as Darwin stressed, this struggle need not be physically violent.[18]

Natural selection pertains to this context. The theory of evolution by natural selection is not very complex – it posits the de facto process of fitter organisms surviving in a given environment. Success for organisms in the struggle will, on average, be a function of their different – and heritable – features. Over time, this will lead to change, according to Darwin:

Let it be borne in mind how infinitely complex and close-fitting are the mutual relations of all organic beings to each other and to their physical conditions of life. Can it, then, be thought improbable, seeing that variations useful to man have undoubtedly occurred, that other variations useful in some way to each being in the great and complex battle of life, should sometimes occur in the course of thousands of generations? If such do occur,

[17] C. Darwin, *The Autobiography of Charles Darwin (1809–1882)*, N. Barlow (ed.) (London: Collins, 1958), 87–88.
[18] Darwin, *Origin of Species*, 62–63.

can we doubt (remembering that many more individuals are born than can possibly survive) that individuals having any advantage, however slight, over others, would have the best chance of surviving and of procreating their kind? On the other hand, we may feel sure that any variation in the least degree injurious would be rigidly destroyed. This preservation of favorable variations and the rejection of injurious variations, I call Natural Selection.[19]

Though Darwin's ideas would be significantly expanded in the coming decades, his key insight that natural selection acts on heritable variation remains a cornerstone of evolution to the present day. As his biologist contemporary and friend Thomas Henry Huxley would comment upon reading *Origin* and appreciating its elegant simplicity, "How very stupid not to have thought of that." The German naturalist Ernst Haeckel, who classified many species and drew an elaborate Tree of Life to show their relationships, celebrated Darwin as Kant's "Newton of the blade of grass."[20]

In 1859, Darwin's remarkable and well-supported theory was still incomplete and in need of more natural mechanisms. For evolutionary change to occur, there must be a steady supply of new variations – and his knowledge of domesticated animals and plants, combined with his systematic study of barnacles over eight years, had strongly convinced Darwin that such variation did occur. Yet he was also convinced that this variation is not correlated to the needs of organisms but rather is "random" with respect to their fitness.[21] Nonetheless, Darwin did not know why and where such variation occurs or how (without being overcome by already-existing forms) a given variation can gain a foothold and persist in a population. Supplying this part of the evolutionary process would have to wait until the early twentieth century, when biologists combined the work on genetics and heredity of nineteenth-century Moravian monk Gregor Mendel with Darwin's theory to achieve a fuller, more adequate theory of evolution.

Though Darwin and Mendel were contemporaries, we have no evidence that Darwin was aware of either Mendel or the implications of Mendel's

---

[19] Darwin, *Origin of Species*, 80–81.

[20] E. Haeckel, *Natürliche Schopfungsgeschicte* (Berlin, 1889), 95.

[21] In biology in general, and evolution in particular, "random" carries the connotation of neither "all outcomes are equally probable" nor "without purpose." There are various chemical features of DNA, for example, that subtly bias mutations to occur more frequently in certain ways and in certain places within a genome. Rather, "random" in evolution simply means that the appearance of mutations is not correlated to the needs of the organism. They are "random" with respect to fitness.

work for his own thinking. Indeed, upon the rediscovery of Mendel's findings in 1900 and the subsequent demonstration that Mendelian principles
apply to a wide variety of organisms, there was debate over whether
Mendelism and Darwinism were compatible.[22] Mendel famously described
character traits that were "recessive" in that they could be masked before
reappearing, unchanged, in later generations. This provided no mechanism
for the appearance of new variation and thus was thought to perhaps be at
odds with Darwinism. Later work would demonstrate that "mutation," the
source of new heritable variation, produced new "alleles," which are variants
of genes. Darwinism and Mendelism were thus compatible, and thus were
combined, and were then recast in the 1930s as "neo-Darwinism" or the
"modern synthesis": the concept depended on changes in allele frequencies
within populations over time.

The modern synthesis continues to guide theorizing and research even as
it has been augmented by other discoveries about evolution in particular and
biology in general. The renewed interest in Mendel in the early 1900s soon
led to the chromosomal theory of inheritance and later to the discovery that
DNA was the hereditary component of chromosomes. The "double helix"
structure of DNA was famously discovered by biologist James Watson and
physicist Francis Crick in 1952, thus providing insight into how it was
duplicated, and also how new alleles might arise through copying errors.[23]
How DNA transmitted its hereditary information to proteins – the processes
of transcription and translation of DNA, via RNA, to the precise sequences of
amino acids that make up proteins – was the key biological question in the
1960s. In the 1970s, DNA sequencing technology began to take shape, and
the complete genome sequences of a few viruses – the smallest genomes
known at the time – were published. The tools of molecular biology – the
ability to manipulate DNA and shape it to desired ends – similarly have their
genesis in this decade. This work accelerated through the 1980s and 1990s,
leading to various "genome projects" that determined the complete DNA
sequence of various organisms and developed additional skills for studying
and manipulating individual genes. The complete genome sequences of
many organisms are now known, including the human genome, and new

---

[22] T. Dobzhansky, "Mendelism, Darwinism, and Evolutionism," *Proceedings of the American Philosophical Society*, 109, no. 4 (1965), 205–215.

[23] J. Watson and F. Crick, "Molecular Structure of Nucleic Acids: A Structure for Deoxyribose Nucleic Acid," *Nature*, 171 (1953), 737–738.

technologies to manipulate DNA with increasing facility and precision continue to be developed.[24]

Against this backdrop of increasing biological knowledge, we have also learned more about evolution. Confirmation that mutations are indeed random with respect to fitness – that is, mutations are not correlated with the needs of the organism – was experimentally demonstrated in the 1930s. Understanding the structure of DNA and its ability to act as a source of both hereditary constancy and novel variation gave additional support to the compatibility of Mendel's and Darwin's ideas. The ability to compare the DNA sequences of different organisms thought to be evolutionarily related on other criteria (anatomy, biogeography, and so on) provided strong evidence for common ancestry. The science of evolutionary devel-opmental biology (so-called evo-devo) demonstrated that development in all animals is accomplished using variations on a common set of early developmental genes that date to before the Cambrian period 600 million years ago.[25]

And yet there were some unanticipated discoveries. Darwinian evolution was strongly "adaptationist" due to its emphasis on natural selection. Natural selection, acting over time, shapes an organism to its environment, and biologists used to thinking in adaptationist terms see the features of the organisms they study in that light. While many features of organisms are under natural selection, a key insight that gained ground in the 1960s was that some features of organisms are *not* under natural selection: they are "neutral" and thus can vary without impacting evolutionary fitness. This is especially true at the level of DNA sequences, the majority of which in most species are not genes (nor regulatory sequences) and have as yet no discern-ible function. For example, estimates of function in mammals indicate that only about 10 percent of mammalian DNA is under natural selection – the approximately 3 percent of the genome that encodes for genes and the approximately 7 percent of DNA that acts in a regulatory role.[26] Indeed,

[24] For a useful and accessible brief history of molecular biology, see J. Tabery, M. Piotrowska, and L. Darden, "Molecular Biology," in E. N. Zalta (ed.), *The Stanford Encyclopedia of Philosophy* (Fall 2019), https://plato.stanford.edu/archives/fall2019/entries/molecular-biology/.

[25] An excellent – though now somewhat dated – introduction to evo-devo is S. B. Carroll, *Endless Forms Most Beautiful* (New York: W. W. Norton, 2005).

[26] K. Lindblad-Toh, M. Garber, O. Zuk, et al., "A High-Resolution Map of Human Evolutionary Constraint Using 29 Mammals," *Nature*, 478 (2011), 476–482.

there is good evidence that the vast majority of mammalian DNA – including human DNA – does not have a sequence-specific function and can vary without consequence. The neutral theory of evolution is thus a strong counterexample to strict Darwinian adaptationism and a good example of how evolutionary biology has moved to a fuller description of life than Darwin envisaged.

Contemporary Philosophical Debates about Design

Although the skeptical Hume, as well as a number of Christian thinkers, had critiqued the logical weaknesses in Paley's argument, it continued to enjoy positive cultural acceptance in the Victorian era because it supported the conviction of a divinely ordained world and social order. For Darwinian evolution to threaten the foundation of that order was a shock to which many were resistant. Nevertheless, the intellectual destruction of Paley's design argument by the facts of Darwinian evolution held only as long as a dichotomy was presumed between divine activity and natural processes, even if those natural processes included chance variations.

Current discussions of the impact of Darwin on the design argument take place on at least two levels – the worldview level and the scientific level – and opinions at both levels vary quite a bit. At the level of philosophical worldview, there are both religious and nonreligious thinkers who accept a clear dichotomy between God and evolution but disagree on how to treat it philosophically. Thinkers who embrace atheism as well as the God/evolution dichotomy believe that Darwin's work solidifies an atheistic interpretation of the universe. Richard Dawkins, the famous Oxford biologist and advocate for atheism, endorses this exact point:

> An atheist before Darwin could have said, following Hume: "I have no explanation for complex biological design. All I know is that God isn't a good explanation, so we must wait and hope that somebody comes up with a better one." I can't help feeling that such a position, though logically sound, would have left one feeling pretty unsatisfied, and that although atheism might have been logically tenable before Darwin, Darwin made it possible to be an intellectually fulfilled atheist.[27]

[27] R. Dawkins, *The Blind Watchmaker* (New York: Norton, 1986), 6.

In other words, those who have always believed that atheism is true in principle can now be "intellectually fulfilled" because Darwinian theory supplies details about how living things can work without God. Of course, since atheism, as the denial of God, is not a comprehensive worldview, it has found its place in modern Western secular culture within the larger framework of philosophical naturalism. The intellectual package of Darwinism and naturalism, which entails atheism, completes a grand narrative in which nature is the exclusive and ultimate reality, physical laws displace God, and science is interpreted as strongly supporting atheism.

By contrast, religious groups that accept the God/evolution dichotomy include the Creation Science movement in its several manifestations and the ID movement, which is a more sophisticated outgrowth of the same underlying religious instincts. But such religious groups take issue with evolution in order to promote their particular view of God's creative activity in nature and combat the spread of atheism in general culture. Since we discussed Creation Science in the previous chapter, let us now take a closer look at the approach to biology called Intelligent Design. According to advocates of ID, there are organic structures that cannot be evolutionary products because the probabilities are staggeringly low that these structures could arise by chance.

Of course, in the *Origin of Species* Darwin argued that chance – both as random variation within a species and as environmental contingency – can and does lead to organisms and their parts being highly adapted to their environment. However, because ID assumes that evolution proceeds by gradual, incremental development from simple to complex structures, ID proponents seek to identify instances of complex, integrated, modular organic structures that could not have developed through the evolutionary process and therefore must have been designed and created as functioning wholes. As supporting evidence, ID proponents list the bacterial flagellum, the blood-clotting cascade, and the eye, among other starting points, for inferring design from the evidence of biology. For biochemist Michael Behe, such structures exhibit "irreducible complexity," although he admits that many other biological structures exhibit reducible complexity.[28] Claiming to "bring design back into biology," mathematician and theologian

---

[28] M. J. Behe, *Darwin's Black Box: The Biochemical Challenge to Evolution* (New York: Free Press, 2006).

William Dembski in 1999 published *Intelligent Design: The Bridge between Science and Theology,* supplementing Behe's idea with his own concept of "specified complexity."[29] We will not pursue these arguments further, but suffice it to say that ID arguments have not gained traction within the mainstream scientific community, where they are widely seen as pseudoscience, after robust critiques by scientists of varying worldviews, including Christians.

One of the more interesting developments in ID discussions revolves around the 2009 book *Signature in the Cell* by Stephen C. Meyer, which claims that "more scientists are becoming interested in the evidence for intelligent design."[30] Meyer assesses the quality of the evidence in biochemistry, molecular biology, and genetics relating to the origin of the information inside cells and concludes that these sciences have reached a dead end following a mechanistic model, such that now the only reasonable *scientific* explanation available is that the information inside of cells must be the product of a transcendent intelligence. This argument presupposes two things: that cellular "information" is not merely *analogous* to information produced by intelligent agents but is in fact information *of the same kind,* and that mechanistic models to explain its origin are at an impasse. Neither premise has mainstream support. Unlike human codes, which are arbitrary in nature,[31] biological information appears to be based, at least in part, on chemical affinities between DNA and amino acids.[32] While a full understanding of the origin of biological information has not yet been achieved (and will not be for the foreseeable future), the features already uncovered are challenging to explain from an ID perspective. Mechanistic explanations for the origin of biological information continue to make fruitful predictions even if the area remains a scientific frontier.[33]

---

[29] W. A. Dembski, *Intelligent Design: The Bridge between Science and Theology* (Downers Grove: InterVarsity Press, 1999).

[30] S. C. Meyer, *Signature in the Cell: DNA and the Evidence for Intelligent Design* (New York: HarperOne, 2009), 436.

[31] R. Isaac, "In Defense of Theistic Evolution," presentation to the ASA at Gordon College July 29, 2018, and "Review of Introduction to Evolutionary Informatics," *Perspectives on Science and Christian Faith,* 69, no. 2 (2017), 99–104.

[32] D. R. Venema, "Seeking a Signature: Essay Book Review of Signature in the Cell," *Perspectives on Science and Christian Faith,* 62, no. 4 (2010), 276–283, and "Intelligent Design, Abiogenesis, and Learning from History: A Reply to Meyer," *Perspectives on Science and Christian Faith,* 63, no. 3 (2011), 183–192.

[33] Venema, "Intelligent Design, Abiogenesis, and Learning from History," 183–192.

Debates over the adequacy of evolutionary science continue between certain sectors of the religious and scientific communities as they each work energetically on the God/evolution dichotomy. No wonder these disputes overshadow more nuanced philosophical renditions of the relationship between God and evolution. Thomas Henry Huxley, who became known as "Darwin's bulldog," realized early on that Darwin's scientific contribution would provoke worldview disputes. Huxley, who was hardly sympathetic to religion, argued that science "commits suicide when it adopts a creed" and that, therefore, in biological science, "the doctrine of Evolution is neither Anti-theistic nor Theistic."[34] His point is that science per se is just science and need not be taken as opposing or favoring either religious or atheistic positions – an insight that sets the stage for our further explorations of how evolution may be incorporated into worldviews that avoid the prevalent dichotomous thinking already surveyed.

While the instincts underlying ID have remained the same over the past few decades – that Darwinian evolution as a product of methodological naturalism cannot explain certain kinds of biological complexity and that ID must position itself as an alternative to mainline biological science – ID has evolved in claiming alliance with other theistic arguments. For example, it is not unusual for ID-inclined websites to mention or link to theistic arguments by Christian author and intellectual C. S. Lewis, who articulated fascinating arguments for God from human reason and from human morality but in fact expressed dislike for design arguments.[35] Aquinas's teleological argument for God has also been favorably mentioned by ID advocates, as has the "fine-tuning argument" that has garnered attention in the last several decades. However, while all of these arguments conclude to an intelligent being beyond this world, they are chronologically prior to and logically independent of the official design argument approach to biology, such that ID per se must be evaluated on its own merits and not by association with other lines of reasoning. Besides, the other arguments, which do not attempt to be arguments from within science, are broadly philosophical and fully compatible with Darwinian evolution.

---

[34] F. Darwin, *The Life and Letters of Charles Darwin* (London: Murray, 1887), 252, 312.

[35] M. L. Peterson, *C. S. Lewis and the Christian Worldview* (New York: Oxford University Press, 2020), 61–62, 109–124. See also C. S. Lewis, *The Problem of Pain* (1940; repr. New York: HarperCollins e-books, 2014), 2–4.

Discussions of Teleology in Current Biology

Besides ongoing philosophical debates over design, there are also current biological discussions of a role for some kind of teleology in the practice of the actual science, but these often fall along a range conditioned by world-view commitments. For instance, in his book *Philosophy of Biology*, Peter Godfrey-Smith argues for "a straightforwardly materialist view of living systems and their evolution," which takes organisms as complex material objects: "In biological systems material things come and go from the world, use energy, change, and give rise to new material things. Evolution by natural selection is one aspect of this great array of physical goings-on."[36] Godfrey-Smith is one of many whose background philosophical commitments lead them to attempt to naturalize teleology in biology and eliminate any of its metaphysical or theological overtones. For those of this mindset, all biological explanation must be mechanistic and not refer to any form of vitalism, mentalism, or a future cause reaching back and influencing the present.[37]

Richard Dawkins famously expressed a strictly materialist view of biology in his modern-day classic, *The Selfish Gene*, which labels all living things "survival machines" subject to purely mechanistic explanation at the level of the gene:

> We are all survival machines for the same kind of replicator – molecules called DNA – but there are many different ways of making a living in the world, and the replicators have built a vast range of machines to exploit them. A monkey is a machine which preserves genes up trees, a fish is a machine which preserves genes in the water; there is even a small worm which preserves genes in German beer mats. DNA works in mysterious ways.[38]

The primary role of genes, then, as "selfish replicators," is to reproduce themselves at all costs. How far the "selfishness" metaphor can be taken is a matter of debate, both as biology and as a worldview assumption – two items we will discuss later.

---

[36] P. Godfrey-Smith, *Philosophy of Biology*, Princeton Foundations of Contemporary Philosophy (Princeton, NJ: Princeton University Press, 2016), 144.

[37] See E. Mayr, *Toward a New Philosophy of Biology: Observations of an Evolutionist* (Cambridge, MA: Harvard University Press, 1988), 38–66.

[38] R. Dawkins, *The Selfish Gene* (Oxford: Oxford University Press, 1976), 26.

Materialist accounts notwithstanding, teleological language referring to function is ubiquitous in contemporary biology – such as the claim that feathers in theropod dinosaurs "served an adaptive function in visual display as opposed to other proposed adaptive functions such as thermoregulation."[39] Those naturalists and materialists who are uncomfortable with teleological language in biology have variously proposed that it be taken as a metaphor or heuristic or shorthand for mechanistic explanation. Currently, there are various attempts to "naturalize" teleology in biology – to bring any such notions under a naturalist philosophical framework – by explaining function purely in terms of natural selection and rejecting any mental or intentional notions.[40] Thus, "teleonaturalism" grounds teleological claims in biology, in facts about organisms and their traits.

Apart from creationists (discussed in Chapter 2) and ID advocates, most biologists and philosophers of biology agree that Darwin deconstructed any Platonic or religious notion of an external *telos* as the basis for explanation in biology. However, some argue that "function talk" in biology cannot be eliminated because of the very nature of the subject matter: living beings. As a historical point, some scholars remind us that Darwin's own evolutionary explanations are indeed teleological. Darwin clearly used the language of final causes to describe the function of biological parts in his Species Notebooks and throughout his career. As philosopher of biology James Lennox observes, Darwin also reflected frequently about the relationship between natural selection and teleology. Rejecting what he calls the "myth" that Darwin replaced teleological explanation of adaptation with nonteleological explanation, Lennox argues that "Darwin was a teleologist" but that he did not conform to the mode of teleology of his time. Rather, "Darwin's explanatory practices conform well … to recent defenses of the teleological character of selection explanations."[41]

Well-regarded biologists such as J. B. S. Haldane and Francisco Ayala note that practicing biologists continue to write as if organisms had goals, almost making the case that teleological language is ineliminable.[42] Knowing, however, that many biologists think it would be ideal if teleological explanations

---

[39] C. C. Dimond, R. J. Cabin, and J. S. Brooks, "Feathers, Dinosaurs, and Behavioral Cues," *BIOS*, 82, no. 3 (2011), 58–63.

[40] See M. Bekoff and C. Allen, "Teleology, Function, Design and the Evolution of Animal Behavior," *Trends in Ecology & Evolution*, 10, no. 6 (1995), 253–255.

[41] J. Lennox, "Darwin Was a Teleologist," *Biology & Philosophy*, 8, no. 4 (1993), 409–421.

[42] F. J. Ayala, "Teleological Explanations in Evolutionary Biology," *Philosophy of Science*, 37, no. 1 (1970), 1–15.

were eliminated, Haldane once quipped that "teleology is like a mistress to a biologist: he cannot live without her but he's unwilling to be seen with her in public."[43] Addressing this dilemma, biologist Colin Pittendrigh coined the term "teleonomy" in the middle of the twentieth century to refer to the obvious goal-directedness of individual organisms according to some built-in lawlike factors, to retain compatibility with physicochemical causality and yet to eliminate connotations of intentional or conscious agency, which is anthropomorphic.[44] Actually, one of Darwin's greatest contributions was to clarify that teleonomic processes pertain to a single individual but are not to be confused with longer-timescale evolutionary changes.[45] Interestingly, the rise of teleonomy connects with the rise of systems thinking and information theory in biology as well.

What does all this have to do with religion? For those religious thinkers who believe that life is a unique quality, teleological language (and even teleonomic language) in biology is a reminder of that special character. For life to resist total reduction by mechanistic explanations throws into question naturalist assumptions about the subject matter of biology as a science and its fit within an overall naturalist worldview while presenting an inviting piece of evidence for interpretation within a religious or theistic worldview. However, the religious viewpoint need not be generally countercultural or at odds with mainline science, as in Creation Science and ID. More sophisticated and more interesting religious viewpoints recognize the seemingly irreducible character of teleological explanation within mainline biological science and offer a theological and metaphysical account of it, such as found in the Thomistic view, which we will return to shortly.

By contrast, those who are committed to naturalizing teleology in biology – and thus define teleonomy naturalistically – believe that teleological explanations in biology are a form of obscurantism aimed at blocking an alternative mechanistic explanation. Biologist K. M. Lagerspetz remarked that "teleological notions are among the main obstacles to theory formation in biology."[46] However, many positions like this are lingering reactions

[43] J. B. S. Haldane in Mayr, *Toward a New Philosophy of Biology*, 63.
[44] C. Pittendrigh, "Adaptation, Natural Selection, and Behavior," in A. Roe and G. G. Simpson (eds.), *Behavior and Evolution* (New Haven, CT: Yale University Press, 1958), 394.
[45] Mayr, *Toward a New Philosophy of Biology*, 38–63.
[46] K. Lagerspetz, "Teleological Explanations and Terms in Biology," *Annals of the Zoological Society Vanamos*, 19 (1959), 1–73.

against early twentieth-century vitalism, and some are conditioned by philosophical commitments to a universe that does not include either purpose or the divine. In any case, it is clear that biologists continue to use teleological language in their explanations of organisms and their parts, displaying almost complete agreement that teleological phrasing of an explanation is not inconsistent with a physicochemical causal explanation. Moreover, to many biologists something seems lacking in explanatory power when any teleology is removed from statements in biology. Consider the change in scientific meaning, not just in heuristic value, between "The sea turtle came ashore to lay her eggs" and "The sea turtle came ashore and laid her eggs." The latter statement is simply not an equivalent logical analysis of the first statement.

Design and Rational Christian Belief

How teleology, or lack of teleology, affects rational explanation of the living world is not the only issue raised between biology and religion. Another question is how the damage done to the design argument by evolution affects the rational status of religious perspectives that affirm design or at least a general teleology. Michael Ghiselin, a philosopher and biologist, expresses the widespread view that Darwinian theory succeeded in "getting rid of teleology and replacing it with a new way of thinking about adaptation."[47] Prior to Darwin, the best explanation for adaptation was to appeal to the argument from design in Paley's natural theology. A major issue then became whether religion – particularly Christianity – was discredited along with the design argument such that a believer could no longer claim rational grounds for religious belief. Let us first explore this issue in its nineteenth-century setting and then in terms of its contemporary context.

Natural theology in Paley's day was approached as an attempt to infer God's existence from aspects of the natural world – a project labeled physicotheology – and the design argument was its main pillar. By identifying the mechanism that produces adaptive complexity, Darwin's theory of natural selection essentially undermined Paley's faulty premise: that such complexity could not occur by chance and therefore must have been intentionally

---

[47] M. Ghiselin forward to C. Darwin, *The Various Contrivances by Which Orchids Are Fertilised by Insects*, 2nd edn. (Chicago: University of Chicago Press, 1984), xiii.

designed. Interestingly, Paley's excessively rationalistic approach to natural theology was already receiving scathing critique from Anglican theologians who contemplated God in relation to nature in a more adequate and more profound manner.

Paley's approach rested on familiar assumed dichotomies – between God and nature, design and chance. However, religious perspectives that rejected the dichotomies as false and misleading could also press toward a more dynamic concept of nature in which God's role is seen as generally supportive and interactive but not always determinative or interventionist. The traditional way of expressing this idea is that God works through "secondary causes," but the precise conception of secondary cause at play here is vitally important. For Paley, secondary causes may be admitted as mechanisms by means of which God creates and orders the world but still remains the ultimate power regulating everything according to a perfect design, which means that God is still the sole cause of all behavior in all natural objects.[48] In Paley's estimation, and probably that of his reading audience, any meaningful autonomy for secondary causes would imply a "chance" element that is inimical to a control-oriented providential design and therefore paves the way to atheism. Thus, the way the nineteenth-century Paley–Darwin controversy was framed, the more chance became recognized as crucial to the evolutionary process, the more a religious view that rejected chance lost its rational credibility. Frankly, by the publication of the Origin, Victorian culture, which had long favored a static design and a set hierarchical order, was already shifting in ways that made it more comfortable with evolutionary ideas.

Although Thomas Henry Huxley stated that evolution devastates teleology "as commonly understood" – and we might say, as narrowly understood – he also believed, in spite of his nonreligious orientation, that evolution reflects a "wider teleology."[49] Hence, Darwinian evolutionary science arguably need not damage teleology as classically conceived but can be one way of exhibiting it. Reinterpreted, the enterprise of natural theology would not so much be inferring God from empirical observations but proposing a theistic worldview framework as more adequate than other worldview frameworks,

[48] Paley, Natural Theology, 418–420.
[49] T. H. Huxley, Lay Sermons, Addresses, and Reviews (London: Macmillan, 1870), 333. See also Darwin, Life and Letters of Charles Darwin, 201.

including scientific naturalism, for making sense of all important aspects of knowledge and human experience.[50] Along these lines, contemporary philosophical discussions of design explanations in relationship to mechanistic explanations have gone in interesting directions. Philosopher of biology Michael Ruse, who holds a naturalist position, maintains that natural selection occurring in a world of natural laws does the work of teleology.[51] However, philosopher Richard Swinburne, a Christian theist, emphasizes that the development of organized complexity over time according to natural laws can reveal a deeper teleology, because the best explanation of the natural laws themselves is that they originate in the mind of God, who is a rational, purposeful agent.[52]

The work of contemporary Christian philosopher Alvin Plantinga has intersected the ongoing discussion of the design argument from a couple of angles. First, his epistemological work on what it means for a belief to be rational is relevant. Given a traditional understanding of natural theology, both religious and nonreligious thinkers saw the Darwinian deconstruction of the design argument as a threat to the rationality of theistic belief. What had always been taken as evident design had become the de facto result of natural selection. However, in the 1980s, Plantinga solidified his critique of natural theology as relying on "classical foundationalism" – or "strong foundationalism" – which is the view requiring that, for a belief to be rational, that belief must be foundational (self-evident or immediately evident to the senses) or inferentially derived from foundational beliefs. Since belief in God is neither self-evident (known to be true by its very meaning) nor evident to the senses (available to ordinary perception), natural theology – say, after the model of Aquinas – seeks to derive the nonfoundational belief in existence of God from foundational beliefs. Thus, his cosmological argument, for example, starts with the observation that a cosmos exists and moves to God's existence.

Plantinga argues that classical foundationalism fails to recognize various kinds of "properly basic" beliefs that are rational to hold without argument

---

[50] M. L. Peterson and M. Ruse, *Science, Evolution, and Religion: A Debate about Atheism and Theism* (New York: Oxford University Press, 2016), 104–114.

[51] M. Ruse, *Darwin and Design: Does Evolution Have a Purpose?* (Cambridge, MA: Harvard University Press, 2003), 6–8. See also Peterson and Ruse, *Science, Evolution, and Religion,* 117–120.

[52] R. Swinburne, *The Existence of God* (New York: Oxford University Press, 1979), 140–141.

from more basic beliefs. His position of "weak foundationalism" recognizes, for example, memory beliefs and beliefs about other minds. His point is that such beliefs are warranted if and only if they are produced by the relevant human cognitive powers, operating normally in the environment for which they were designed. He refers to theologian John Calvin's idea of the *sensus divinitatis* – the sense of the divine or instinctive inclination to believe in God – as one more faculty in the panoply of our noetic equipment. Hence, according to what is called Plantinga's "Reformed epistemology," it can be (under the conditions mentioned) rational to believe in God without argument, although arguments can be an additional support.[53]

A second angle of approach in Plantinga's engagement with science is his argument that the scientific ideal of methodological naturalism is untenable. The ideal of a worldview neutral method of investigation is naively mistaken assuming that scientists do not allow their background beliefs to influence their conclusions. Therefore, "the Christian community can't automatically take the word of the scientific experts; sometimes what the experts say presupposes a philosophical or religious stance quite opposed to that of Christian theism."[54] In biology specifically, Plantinga contends that many scientists take randomness and chance in evolution to imply that human beings are purely accidental, not *intended* by God. For example, famous zoologist George Gaylord Simpson asserted that "man is the result of a purposeless and natural process that did not have him in mind."[55] However, according to Plantinga, Simpson's naturalist assumptions led him to abandon neutral methodological naturalism and inflate chance from the biological level to the ultimate metaphysical level. By contrast, claims Plantinga, theists would not make this jump because they could acknowledge the evidence for evolutionary chance while still believing humanity to be purposed by God.[56] Disagreement with Plantinga's proposal of "theistic science" – into which theists can bring all that they know to bear on their science – has been expressed by both Christian believers and nonbelievers who think that the ideal of methodological naturalism is indeed the core of

[53] A. Plantinga, "The Reformed Objection to Natural Theology," *Christian Scholar's Review*, 11 (1982), 187–198.
[54] A. Plantinga, "Science: Augustinian or Duhemian?," *Faith and Philosophy*, 13, no. 3 (1996), 369.
[55] G. G. Simpson, *The Meaning of Evolution*, rev. edn. (New Haven, CT: Yale University Press, 1967), 344–345.
[56] Plantinga, "Science: Augustinian or Duhemian?," 372–373.

science. Catholic philosopher of science Ernan McMullen responded that "methodological naturalism ... lays down which sort of study qualifies as scientific. ... Scientists have to proceed in this way."[57] Eugenie Scott – who is a scientist, self-avowed secular humanist, and former director of the National Center for Science Education – has stated a similar view: "In doing science, one has to proceed as if there were no supernatural interference in the operations of nature."[58]

Moderating Positions

In addition to the diametrically opposed positions – that organic complexity must result from either divine intervention or evolutionary processes – there is a range of intermediate or moderating positions, falling mostly under the rubric "theistic evolution." Since much of the polarization flows from dichotomies – God/nature, divine action/natural process, purpose/chance – views that avoid these dichotomies find various avenues for combining scientific facts and religious or quasi-religious interpretations of the facts. Some narrower or more parochial versions of theistic evolution admit some but not all of the scientific facts and leave conceptual space for some degree of interventionism – and thus end up as one rendition or another of "old Earth" creationism, which modifies Young Earth Creationism. However, what we want to explore are views that more straightforwardly accept well-confirmed science, including evolution, while still offering a framework that involves a divine or nonmaterial element.

An example of a view that does not hedge on science would be Henri Bergson's notion of "creative evolution" (see Chapter 2), which posits the idea of the *élan vital*, a creative impulse driving change in the universe. This may seem like a form of theistic evolution, but a creative life force at the heart of reality does not actually qualify as standard or classical theism, with its omnipotent, omniscient God. Instead, Bergson's views are much closer to process theism, inspired by philosopher Alfred North Whitehead's meta-physical vision, which posits that there is a universal principle of creativity and integration working toward a synthesis of all human knowledge and

[57] E. McMullin, "Plantinga's Defense of Special Creation," in R. T. Pennock (ed.), *Intelligent Design Creationism and Its Critics* (Cambridge, MA: MIT Press, 2001), 168.

[58] E. C. Scott, "Science, Religion, and Evolution," National Center for Science Education (June 18, 2004), https://ncse.ngo/science-religion-and-evolution.

experience. Interestingly, several Hindu thinkers, who believe the Vedantic principle that Brahman is the hidden inner essence of all things, have resonated with Bergson's idea that some sort of consciousness underlies evolutionary change.[59]

Arguably, the most classic and enduring version of theistic evolution stems from the thought of St. Thomas Aquinas and continues in the broad Thomistic tradition. Aquinas's master principle in the *Summa Theologica* is that faith and reason cannot actually contradict each other, and that all truths about reality are harmonious and unified in a theistic universe, regardless of whether there are apparent incongruities and conflicts that we have yet to resolve in practice. Although Aquinas comes at the end of an era in which a hierarchical and essentialist view of the universe was held, his metaphysical and theological understanding of primary and secondary causality affords theism an important connection to the world of science. Rather than look for gaps and incompletions in science in order to make room for God, Thomistic metaphysics envisions God as the primary cause – the infinite agent that initiates and sustains all finite existence, creating what Aquinas calls secondary causal agents with their distinctive powers and properties. Applied to the evolutionary character of the biological realm, this means that all of the physical conditions that make for life and its diversification are due to capabilities of the chemicals, genes, and other factors that underlie evolution.

Thomistic theism has been interpreted by some as a control-oriented or deterministic theism, but Aquinas actually recognized a role in creation for nondetermined contingencies – chance events.[60] For him, human choice must in important ways be nondetermined in its outcomes. As Aquinas indicates, the creation, while orderly, also includes chance (*accidens*) and luck (*fortuna*) – in other words, nondetermined contingencies. The role of chance in evolutionary theory – both as random variation and as environmental happenstance – is not a threat for this understanding of theism, which sees God as empowering creatures with their own "functional integrity" and giving them space to act. In this same vein, Charles Kingsley, a nineteenth-century Anglican minister and professor who was well aware of Darwin's work, rejected any false dichotomies between God and natural causes and

---

[59] K. N. Aiyer, *Professor Bergson and the Hindu Vedanta* (Vasanta Press, 1910), 36–37.

[60] Aquinas, *Summa Contra Gentiles*, 3.1.72–77.

envisioned evolution as a process in which creatures with God-given powers and potentialities could dynamically develop in relation to environmental changes. Kingsley perceptively remarked, "We knew of old that God was so wise that He could make all things: but behold, He is so much wiser than even that, that He can make all things make themselves."[61]

Thus, a Thomistic version of theistic evolution is an illustration of an approach providing theological grounds for the idea that biological organization – the outcome of variation, inheritance, and selection – could emerge over time under the providence of God. Although exploration of what providence could mean in this context lies ahead, it is worth noting a statement on this topic by physicist-theologian John Polkinghorne:

> The actual balance between chance and necessity, contingency and potentiality, which we perceive seems to me to be consistent with the will of a patient and subtle Creator, content to achieve his purposes through the unfolding of process and accepting thereby a measure of the vulnerability and precariousness which always characterize the gift of freedom by love.[62]

Polkinghorne suggests that God's interaction with the world is not based on a predetermined script but is rather an improvisation between God and creatures. Such a statement correlates with what is known as open theism, which has arisen in the last few decades and will be considered in Chapter 5.

---

[61] C. Kingsley, preface to *Westminster Sermons* (London: Macmillan, 1881), xxv.
[62] J. Polkinghorne, *One World: The Interaction of Science and Theology* (West Conshohocken, PA: Templeton Press, 2007), 82.

4   Biology and the Problem of Natural Evil

The problem of evil has perennially been the strongest objection to religious faith among nonbelievers as well as a major source of crises of faith among believers. The problem centers on the alleged incompatibility between God and evil of all kinds in the world that he supposedly created and governs. On the face of it, this incompatibility seems to support agnosticism or atheism and thus creates difficulties for religion, particularly theistic religion. Since the mid-twentieth century, formulations of the problem – as well as responses to it – have become increasingly technical due to the rise of the analytic philosophy of religion. Moral evil was a salient point of debate, and still is, but theistic replies revolving around the concept of free will have come to be considered quite effective.

In recent decades, theistic critics have increasingly focused on the problem of natural evil as perhaps a more difficult challenge for theistic belief, particularly when formulations of the problem are shaped by evolutionary understanding. In the eighteenth century, skeptic David Hume pressed the problem of natural evil, enumerating various maladies in humans from "intestine stone and ulcer" to death itself. He was also well aware that animals share in the ravages of nature: "The stronger prey upon the weaker and keep them in perpetual terror and anxiety."[1] Yet it was left to Darwin in the nineteenth century to explain scientifically that suffering is endemic to the animal world due to natural selection. Although the phenomena of physical pain and suffering have motivated various renditions of the problem of natural evil, our focus is on the problem as shaped by Darwinian evolutionary biology and religious replies to it. Some religious responses still

---

[1] D. Hume, *Dialogues Concerning Natural Religion* (1779), 10.9.

link all natural evils to moral evil in the human realm, but we will focus mainly on what we label the Darwinian problem of animal pain.

Religion and the Problem of Evil

All religions face problems of evil that are structured from within their own distinctive belief systems and require a response based on their own conceptual resources. For example, Hinduism asserts that the law of karma gives each creature its just due; Buddhism teaches that existence simply is suffering; and Zoroastrian dualism holds that pain and evil stem from warfare between good and evil cosmic powers. Yet theistic religions – based on the idea that God is omnipotent, omniscient, and wholly good – face the most difficult formulations of the problem of evil: the divine attributes are so lofty that they generate the highest expectations, and yet the expectations for what this God would do in the created world seem not to be met. Citing Epicurus, Hume makes the point:

> Is [God] willing to prevent evil, but not able? Then he is impotent. Is he able, but not willing? Then he is malevolent. Is he both able and willing? Whence then is evil?[2]

In the discussion to follow, we must look closely at how this point plays out in relation to theism and biological science.

In philosophy generally, a problem is an argument that has credible premises but that leads to a conclusion that one does not want to accept. The problem of evil for theists is how to avoid an atheistic conclusion by rebutting one or more premises in the argument. During most of the twentieth century, critics posed the problem as a *logical argument* – basically, that God and evil cannot coexist; it is impossible for them to coexist. Since theism affirms the existence of both God and evil in his world, the argument was advanced by critics as revealing a logical inconsistency in theistic belief. Such was the prevailing view in professional philosophy, as expressed by atheist philosopher J. L. Mackie:

> In its simplest form the problem is this: God is omnipotent; God is wholly good; and yet evil exists. There seems to be some contradiction between these

---

[2] Hume, *Dialogues*, 10.5.

three propositions, so that if any two of them were true the third would be false. But at the same time all three are essential parts of most theological positions: the theologian, it seems, at once *must* adhere and *cannot consistently* adhere to all three.[3]

Obviously, if it can be shown that theism is internally inconsistent and thus irrational, it would be the ideal coup for nontheistic critics.

In the early 1970s, however, Christian philosopher Alvin Plantinga articulated a highly sophisticated version of the "free will defense," which argued that it is logically possible that God would give free will to finite rational persons with the risk that they might misuse it to bring about moral evil. Holding a conception of libertarian free will, Plantinga argued that, if creatures misused their freedom of choice, then moral evil would exist in God's created universe. Since libertarian free will is incompatible with any form of determinism, whether divine power or natural forces, the critic is mistaken that logic dictates that there would be no moral evil in a world created by an omnipotent and wholly good God. Therefore, the critic's case that God and evil cannot coexist fails. In less than a decade, the success of Plantinga's argument was widely recognized by theists as well as their critics, and so the strategy for articulating a problem of evil for theism migrated away from the logical problem. While critics recognized that moral evil supplies some evidence against theistic belief, they also realized that natural evil appeared to be important evidence and had the added virtue of not being obviously subject to the free will defense. The evaluation of theistic responses specifically to the problem of natural evil in the nonhuman living world will occupy the rest of this chapter.

The Problem of Natural Evil

Since the 1980s, the evidential problem of evil formulated by atheist philosopher William Rowe has been the fountainhead of most discussions on this topic. Rather than pose the problem for theism as one of logical inconsistency, Rowe's argument cites natural evil as "evidence" against the credibility of theism.[4] Although Rowe's argument went through several iterations over a couple of decades, the basic argument goes as follows:

---

[3] J. L. Mackie, "Evil and Omnipotence," *Mind*, 64, no. 254 (1955), 200.
[4] W. L. Rowe, "The Problem of Evil and Some Varieties of Atheism," *American Philosophical Quarterly*, 16, no. 4 (1979), 335–341.

1. There exist evils in our world that are not necessary to the existence of any greater good or to the prevention of any evil equally bad or worse.
2. An omnipotent, omniscient, wholly good God would not allow any evil unless it is necessary to the existence of a greater good or the prevention of an evil equally bad or worse.
3. Therefore, an omnipotent, omniscient, wholly good God does not exist.

Premise 1 is a statement of the evidence, a factual claim to the effect that there appear to be evils that are not necessary for either greater goods or the prevention of evils. For shorthand, let us call these *gratuitous evils*. Premise 2 is a theological claim – essentially, that God's stated attributes entail that he would not allow gratuitous evil. Since premise 1 is a factual claim that is inductively grounded, it is not logically certain but is rather to some degree probable. Rowe thinks that the factual premise is very probable, more probable than not, which would make the atheistic conclusion more probable than not.

As evidence for premise 1, Rowe originally described an instance of natural evil – a fawn trapped in a forest fire and suffering in torment for days before dying. For Rowe, this approach shows that the problem of evil is not properly cast as a logical problem and that it is not exclusively about human pain and suffering. In later discussions, the fawn example was labeled "the case of Bambi," and another example of specifically human suffering was added – the beating, rape, and murder of a five-year-old girl by her mother's drunken boyfriend – which became called "the case of Sue." According to Rowe, these evils – one natural and the other moral – are examples of uncountably many instances of evil occurring daily in our world. Our exploration here deals with evils occurring in the nonhuman animal world – the problem of creaturely suffering – but recognizes its relevance to human suffering because humans are a part of the animal world.

Since the logic of the argument is correct, an effective theistic response to the evidential argument must sufficiently lower the credibility of one or more of the premises. Historically, most theistic responses have targeted the factual premise – which for our study asserts the claim that there is widespread, intense animal pain and suffering. Two broad approaches for rebutting the factual premise have emerged – defense and theodicy. A *defense* is a reason why we cannot be in a position to know or reasonably believe that the factual premise is true, whereas a *theodicy* is a reason why we can know or justifiably believe the factual premise to be false.

The position that we cannot know the premise to be true has been called the "skeptical theist defense" – because it targets the epistemic standing of the critic or any other person forming a belief about the existence of gratuitous evil. Stephen Wykstra, Michael Bergmann, and others have argued that human cognitive capacities are finite and thus not capable of reliably knowing all connections between good and evil; thus humans are not capable of knowing whether or not a particular evil is connected to some greater good. By contrast, God's infinite wisdom would know all connections between evils and greater goods. The defense, then, supposedly removes any rational grounds Rowe has for judging the truth or probability of the factual premise – because finite persons simply do not have the capacity to know whether there are justifying goods connected to the evils we witness. Rowe responded to the skeptical point by saying that there are so many cases of "apparently gratuitous evil" that surely we cannot be mistaken about all of them, such that it is reasonable to believe that at least some are genuinely gratuitous. Many theists have also rejected skeptical theism because the classical Christian doctrine of humanity entails that God gave humanity reliable cognitive and moral powers that resemble in finite ways God's own infinite powers – thus making highly suspect any blanket claim that human powers are systematically mistaken in these matters.[5] The debate over the effectiveness of skeptical theism continues, but we will not pursue it here.[6]

Historically, of course, theodicy rather than defense has been the most frequent strategy for responding to the evidential argument by maintaining that its factual premise is false – in other words, arguing that that there are no gratuitous evils because all evils are necessary to some greater good. Greater-good theodicies that address natural evil specifically often cite the lawlike operation of the physical world as necessary to a stable, flourishing existence – gravity allows traction and motion; cell replication allows growth and repair.[7] Nonetheless, as this theodicy explains, the same

---

[5] M. L. Peterson and M. Ruse, *Science, Evolution, and Religion: A Debate about Atheism and Theism* (New York: Oxford University Press, 2016), 196–206; M. L. Peterson, "Christian Theism and the Evidential Argument from Evil," in D. Werther and M. D. Linville (eds.), *Philosophy and the Christian Worldview* (New York: Continuum, 2012), 175–195.

[6] T. Dougherty and J. P. McBrayer (eds.), *Skeptical Theism: New Essays* (New York: Oxford University Press, 2014).

[7] R. Swinburne, "Natural Evil," *American Philosophical Quarterly*, 15, no. 4 (1978), 295–301.

justifiable natural laws sometimes bring about events that conflict with human wishes and agendas – from accidentally falling off a cliff, which allows gravity to bring death, to faulty cell replication, which can cause cancer. Although evils arising from the regular operations of the physical world are part of the debate, we will direct our attention even more specifically to the problem of animal pain: there appears to be gratuitous, widespread, intense animal pain in our world; if the theistic God exists, then there would not be gratuitous animal pain; therefore, probably the theistic God does not exist.

Many theistic thinkers and their critics believe that an adequate theodicy in general – and theodicy for the problem of creaturely suffering in particular – must rebut the factual premise regarding the gratuity or pointlessness of animal suffering. Prominent strategies for straightforwardly rejecting the factual premise include Cartesian, Augustinian, Irenaean, and process theodicies, which we will analyze later with particular interest in how well they comport with biological information. In the end, we evaluate how the evidence of animal pain impacts the overall rational choice of worldviews.

Cartesian Responses

A whole range of Cartesian responses to the problem of animal pain maintains either that animals do not suffer or that their suffering is not morally significant. Such responses rely on René Descartes's argument that animals do not suffer because they do not have minds. In the seventeenth century, Descartes endorsed a version of metaphysical dualism between "mind" (which he called *res cogitans*, or thinking substance) and "matter" (which he called *res extensa*, or extended substance). According to Descartes, a human being is a mind that is related to a material body, but an animal is simply a material body without a mind and thus without the ability to think, much like a robot operating mechanically, "an automaton." For dedicated Cartesians, the machine analogy – that the body may be regarded as "like a machine" – applies to animals because their complex movements can be explained in terms of physiological laws. By contrast, Descartes held that the distinctive feature of human persons is "thought," which covers "everything that is in us in such a way that we are immediately conscious of it." Thus, for

Descartes, "all operations of the will, the intellect, the imagination, and the senses are thoughts."[8]

Hence, a Cartesian theodicy for the evidential problem of animal pain is a Cartesian explanation of why animals do not in fact feel pain – which is tantamount to the denial of the factual premise. This is not to deny Descartes's recognition that we learn from an early age to ascribe pain to animals because their external behaviors appear to signal pain, although we have no direct, reliable access to anything like their minds or souls. Later Cartesians – such as Antoine Arnauld and Nicholas Malebranch – applied this view precisely to the problem of theodicy, arguing that divine goodness morally requires that animals not really feel the pain that their behaviors seemed to signal. God's creation of a Cartesian dualistic universe, therefore, fulfills that moral requirement.[9] While Norman Kemp Smith has called it a "monstrous thesis" that "animals are totally without feeling" and possess only those characteristics found in the rest of mechanistic nature, the classic Cartesian position is a starting point for exploring contemporary neo-Cartesian views that can be applied to theodicy for apparent animal pain and suffering.[10]

Among Descartes scholars, various debates concern whether not having a mind necessarily implies not having feeling or sensation, which, as even Descartes admits, could simply depend on some physical organ. For example, some interpretations, rooted in certain Cartesian texts, distinguish between levels of conscious awareness – perhaps sensation without thought – which leads to the idea that lower-level sensations, stemming from some corporeal organ, are still not mentally interpreted and given significance by higher-level rational, reflective processing. On this neo-Cartesian view, then, animals are not merely unconscious automata: they are alive, awake, and have sensations, but they do not have a unified, inner, subjective, phenomenal sense of the pain. In terms of modern discussions of consciousness, they are perhaps best viewed as phenomenally unconscious zombies.[11] Other neo-

[8] R. Descartes, *Meditations on First Philosophy: With Selections from the Objections and Replies*, Latin-English edn. (New York: Cambridge University Press, 2012), 151.

[9] M. J. Murray, *Nature Red in Tooth and Claw: Theism and the Problem of Animal Suffering* (New York: Oxford University Press, 2008), 6.

[10] N. K. Smith, *New Studies in the Philosophy of Descartes* (London: Macmillan, 1952), 136, 140. The fuller topic is discussed in J. Cottingham, "'A Brute to the Brutes?': Descartes' Treatment of Animals," *Philosophy*, 53, no. 206 (1978), 551–559.

[11] J. J. Lynch, "God, Animals, and Zombies," *Ágora Papeles de Filosofía*, 30, no. 2 (2011), 13–25.

Cartesian proposals allow that animals may have pain states but lack the rational, reflective capacity to regard those states as undesirable, much like the reports of human individuals who have undergone a prefrontal lobotomy, for example.[12]

C. S. Lewis, an influential Christian thinker and author of *The Problem of Pain*, takes a neo-Cartesian view in a chapter dedicated specifically to the problem of why God permits animal pain. After explaining how pain may be a catalyst for moral and spiritual improvement in human persons, he admits that nonrational animals are not subject to the same explanation. Essentially, Lewis takes the position that the inner states of beasts – including what appear to be pain states – are completely unknowable to us. In any case, as he argues, a sentient but nonhuman creature has no enduring, conscious self or soul to recognize reflectively "I am in pain" or to experience and unify a succession of pain states as "suffering."[13] Clearly, one of Lewis's reasons for this position is that to ascribe selfhood or soulhood to animals might blur the distinction with humans, which are animals made in the image of God with rational, moral, and spiritual capacities that make them unique and special. Thus, Lewis gives no support to notions of animal salvation:

> The real difficulty about supposing most animals to be immortal is that immortality has almost no meaning for a creature which is not "conscious" in the sense explained above. If the life of a newt is merely a succession of sensations, what should we mean by saying that God may recall to life the newt that died today? It would not recognize itself as the same newt; the pleasant sensations of any other newt that lived after its death would be just as much, or just as little, a recompense for its earthly sufferings (if any) as those of its resurrected – I was going to say "self," but the whole point is that the newt probably has no self.[14]

He, nonetheless, speculatively muses that in domesticating higher animals we impart something of ourselves to them, such that they may indirectly have an afterlife as God grants immortality to their masters.[15]

[12] See M. J. Murray and G. Ross, "Neo-Cartesianism and the Problem of Animal Suffering," *Faith and Philosophy*, 23, no. 2 (2006), 177.

[13] C. S. Lewis, *The Problem of Pain* (1940; repr. New York: HarperCollins e-books, 2014), 135–137 ff.

[14] Lewis, *Problem of Pain*, 142.

[15] See also T. Dougherty, *The Problem of Animal Pain: A Theodicy for All Creatures Great and Small* (New York: Palgrave Macmillan, 2014).

Criticisms of the neo-Cartesian principle that animals do not experience pain are abundant and threaten to undermine neo-Cartesian theodicies that attempt to justify God's permission of "merely apparent" animal pain. For one thing, common sense suggests that animals surely experience some unpleasantness or pain because they react and display avoidance behaviors similar to those of humans. For another thing, several areas of science reinforce common sense. For example, neuroscience reveals similarities of neural structure, from which we might naturally conclude that animals suffer in much the same ways humans do. Clearly, mammal brains are constructed similarly – brainstem, limbic system, cerebellum, and cerebral cortex – with the major differences between species being size and complexity of the cortex. Furthermore, ethology, the study of animal behavior, has found it profitable to attribute cognitive states to animals in explaining their behavior. Last, attachment theory in social psychology has expanded from humans to other animals by incorporating biological insights into how emotions like love, anger, and sorrow help various species survive and function.[16] This accumulated information makes it extremely difficult to accept the neo-Cartesian view that animals do not suffer in ways comparable to our own.

Darwin himself argues in *The Expression of the Emotions in Man and Animals* that animals feel emotions.[17] Today, extensive studies confirm and refine Darwin's thought, including the body of research by noted primatologist Franz de Waal on the emotional states and social dynamics of the great apes, which express feelings in much the same way as humans.[18] De Waal asks, "Are we open-minded enough to assume that other species have a mental life?"[19] Although his writings contain many amazing reports of primate behavior resembling the human, his retelling of a 1996 news story about a captive female gorilla named Binti is particularly compelling. It begins when a three-year-old boy fell 18 feet into the primate exhibit at Chicago's Brookfield Zoo:

---

[16] See, for example, B. King, *How Animals Grieve* (Chicago: Chicago University Press, 2013); M. Bekoff, "Animal Emotions: Exploring Passionate Natures," *BioScience*, 50, no. 10 (2000), 861–870.

[17] C. Darwin, *The Expression of the Emotions in Man and Animals* (Chicago: University of Chicago Press, 1872).

[18] F. de Waal, *Mama's Last Hug: Animal Emotions and What They Tell Us about Ourselves* (New York: W. W. Norton, 2020).

[19] F. de Waal, *Are We Smart Enough to Know How Smart Animals Are?* (New York: W. W. Norton, 2016), 34.

Reacting immediately, Binti scooped up the boy and carried him to safety. She sat down on a log in a stream, cradling the boy in her lap, giving him a few gentle back pats before taking him to the waiting zoo staff. This simple act of sympathy, captured on video and shown around the world, touched many hearts, and Binti was hailed as a heroine. It was the first time in U.S. history that an ape figured in the speeches of leading politicians, who held her up as a model of compassion.[20]

Barbara King, an anthropologist specializing in grief in animals, indicates that a mother gorilla will often carry the corpse of her dead baby for weeks and that elephants will often visit the bones of their relatives. According to King, we are amassing a large database in the peer-reviewed scientific literature, ranging from chimpanzees and elephants to dolphins, giraffes, and birds – data that increasingly confirms the theory that animals grieve.[21] In fact, the scientific evidence for grief and mourning in other species has led the journal *Philosophical Transactions of the Royal Society of London* to suggest that there should be a new field of study labeled "evolutionary thanatology."[22]

However, voicing continuing support for the neo-Cartesian line, Peter Harrison, a philosopher of science and religion, argues that there are good reasons, scientific and otherwise, to believe that animals do not experience pain – a claim contradicting Peter Singer's statement that there are good reasons for thinking animals do feel pain.[23] For example, similarity in neurological structure, argues Harrison, is not conclusive proof of similarity in pain sensation. Citing Thomas Nagel's famous essay, "What Is It Like to Be a Bat?," Harrison claims that we cannot know the subjective experience of other kinds of sentient creatures.[24] Along the same lines, David DeGrazia, a philosopher of cognitive science, observes that the attribution of cognitive states (such as beliefs and intentions) may be utilitarian rather than

[20] F. de Waal, *Our Inner Ape* (New York: Riverhead Books, 2005), 3.

[21] King, *How Animals Grieve*.

[22] J. R. Anderson, D. Biro, and P. Pettitt, "Evolutionary Thanatology," *Philosophical Transactions of the Royal Society, Series B*, 373 (2018), DOI: https://doi.org/10.1098/rstb.2017 .0262.

[23] P. Harrison, "Do Animals Feel Pain?," *Philosophy*, 66, no. 255 (1991), 25–40; P. Singer, *Animal Liberation* (London: Cape, 1976), 16. See also J. J. Lynch's "Harrison and Hick on God and Animal Pain," *Sophia*, 33, no. 3 (1994), 62–73, and "Is Animal Pain Conscious?," *Between the Species*, 10, no. 1 (1994), 1–7.

[24] Harrison, "Do Animals Feel Pain?," 31–32.

ontological, practically useful but not about reality.[25] Utilitarian attribution is heuristic or operational, making no claim about the actual inner states of animals while still predicting a number of behaviors. Interestingly, various episodes in the history of science are based on false theories predicting many correct outcomes – Ptolemaic astronomy, for example, correctly predicted a lot of data points. By contrast, ontological attribution commits to the reality of the inner states of animals as causes of their behaviors.

Evolutionary psychologist Gordon Gallup Jr. cautions that we should restrict the attribution of inner cognitive and emotional states to animals that have "second-order" mental states – that is, self-awareness. Gallup states:

> Either you are aware of being aware or you are unaware of being aware, and the latter is tantamount to being unconscious. The sleepwalker is sufficiently aware to navigate and avoid colliding with obstacles, but unaware of being aware . . .. If a species fails to behave in ways that suggest it is aware of its own existence, then why should we assume it is aware of what it is doing?[26]

In 1970, Gallup developed the mirror self-recognition (MSR) test to identify animals with self-consciousness. Some animals with red dots placed surreptitiously on their foreheads look in a mirror and immediately touch the red dot, which Gallup claims as significant evidence for self-consciousness. Humanoid primates passed the early experiments, but the list of animals with positive results has grown to include, at a minimum, dolphins, elephants, and magpies.[27] By implication, then, we should be careful in attributing emotions and pain to animals that fail or could be assumed to fail MSR.

Augustinian Theodicy and Animal Pain

Historically, most Christian theodicies do not rely on Cartesian thinking about animal pain but rather focus on other issues related to the human condition. For 1,500 years in the West, St. Augustine's views have formed the dominant template for Christian theodicy, which includes a distinctive

[25] D. DeGrazia, *Taking Animals Seriously: Mental Life and Moral Status* (Cambridge University Press, 1996), 87.

[26] G. G. Gallup Jr., "Do Minds Exist in Species Other than Our Own?" *Neuroscience and Biobehavioral Reviews*, 9, no. 4 (1985), 638.

[27] K. Andrews, "Animal Cognition," in E. N. Zalta (ed.), *The Stanford Encyclopedia of Philosophy* (Summer 2016), https://plato.stanford.edu/archives/sum2016/entries/cognition-animal/.

explanation of animal pain within its broad scope. For Augustine, a perfect God created a perfect world "out of nothing" (*ex nihilo*) – a physical environment and human beings situated within it. Augustine's neo-Platonic background led him to apply Plotinus's "principle of plenitude" to creation: that God would want creation to be rich, diverse, and full, exhibiting an amazing variety of creatures that would differ in terms of beauty, strength, and intelligence. The creation constitutes a natural hierarchy in which human beings are the crowning achievement, made "in God's image" (*in imago Dei*). God pronounced the whole creation "good" – and God pronounced humanity "very good." The perfection of the world is not absolute perfection, which only God possesses, but a finite perfection of the creature as God's handiwork. From the early twentieth century to the present, Protestant fundamentalist theodicies, which are one extreme on the spectrum of Augustinian-type theodicies, adamantly insist on a "young Earth" and the instantaneous creation of all species in unchangeable forms, positions that have been shown to be blatantly false. Clearly, Augustine himself did not hold such views; he believed that the book of Genesis could be read figuratively on matters for which we already have factual knowledge in order to extract the fundamental theological truths.[28]

For Augustine, the flaw in creation occurred with the human fall into sin. Adam, the first human, misused God's gift of free will and fell into a state of sin and rebellion against God. The "mystery of free will" for Augustine concerns how a finitely perfect and good being could choose to go wrong and to turn its back on the Creator. Although Augustine claimed that the mutability and changeability of something made out of nothing could perhaps account for the free creature's choice to sin, the creature is, nevertheless, spiritually and morally responsible for sin, not God. The consequences of the original human sin were not limited to human alienation from God but extended throughout creation, damaging its pristine order and beauty. Struggle, pain, predation, and death in the nonhuman animal realm were part of this damage.

Many Augustinian thinkers assert that Adam and the rest of creation were "corrupted" by angelic beings with free will – that indeed their fall led to rebellion in the rest of God's creation. Nowhere has this view been more powerfully presented than in *Paradise Lost*, the epic poem by the great

---

[28] Augustine, *The Literal Meaning of Genesis*, 1.19.

seventeenth-century English poet John Milton, who made Satan the central figure in the Fall. But by the early nineteenth century, evidence was mounting that animals predated humans, making the idea that animal pain resulted from either a human or an angelic Fall increasingly difficult to defend. It is no wonder that many Christian theodicists and apologists since Darwin have realized that they must account for animal pain in some other fashion – indeed, that they must construe the evolutionary process as God's way of structuring the physical creation.

Yet to the present day, Augustinian themes in theodicy remain deeply ingrained in much Christian thinking in the West, making it difficult to eliminate the notion that natural evil is the result of moral evil. This fallen world is therefore in need of God's redemptive activity to restore relationship with humans and concomitantly restore the original order and beauty of the natural world. Thus Augustinian theodicy is framed around the familiar V-shaped pattern of creation–fall–redemption. As redemption of the whole creation is accomplished in the eschatological future, animal pain, while a secondary matter for Augustinian theodicy, will be eliminated along with human pain. Here we have largely a causal explanation of why past sin led to all suffering, including animal pain, coupled with the promise of an unimaginably wonderful, pain-free future for all redeemed creatures, including animals.

Augustinian free will–type theodicies are subject to various criticisms – say, regarding whether the restoration of the aesthetic beauty of the created universe is morally sufficient compensation or justification for the world's history of pain and suffering, including animal pain and suffering, that supposedly resulted from the Fall. Obviously, scientific evidence puts enormous rational pressure on an Augustinian theodicy for biological pain and suffering. We know, for example, that paleoanthropology, genetics, and other fields indicate that scientifically there could not have been a single primeval couple, Adam and Eve, such that they could have been the cause of the pervasive pain in the animal world. Instead, we learn from science that the present genetic diversity of modern humans is consistent with an average effective population size in the tens of thousands over the last several hundred thousand years, reaching back to a time prior to the earliest examples of our species in the fossil record.[29]

[29] H. Li and R. Durbin, "Inference of Human Population History from Individual Whole-Genome Sequences," *Nature*, 475, no. 7357 (2011), 493–496. Published 2011 July 13, DOI:

Yet perhaps the deepest level of critique of Augustinian thinking on natural evil pertains specifically to the origin and role of pain in the living world. We know from modern evolutionary science that pain is not an intrusion into a previously painless biological realm but is rather part of its warp and woof. In the struggle for survival, carnage, predation, pain, suffering, and death are inextricably linked to the production of more fit organisms that will survive and reproduce. Darwin was struck by this condition of nature, writing in 1856 to his friend Joseph Hooker, "What a book a devil's chaplain might write on the clumsy, wasteful, blundering, low and horridly cruel works of nature."[30] Although Darwin's remark reflects his increasing doubt that this could be "a perfect world" designed by a perfect God (an Augustinian idea that Darwin encountered in theologians in his culture), he usually allowed Thomas Henry Huxley to press the point.

In 1860, Darwin commented to his Christian friend Asa Gray, a Harvard biologist, how natural evils seem to count against the existence of a good God who created the world:

> I own that I cannot see as plainly as others do, and as I should wish to do, evidence of design and beneficence on all sides of us. There seems to me too much misery in the world. I cannot persuade myself that a beneficent and omnipotent God would have designedly created the Ichneumonidae with the express intention of their feeding within the living bodies of Caterpillars, or that a cat should play with mice.[31]

Apparently, it is not just currently extant species that are tormented by parasites. In his book, *Animal Suffering and the Darwinian Problem of Evil*, theologian John Schneider reported discoveries of fossilized amber from the late Cretaceous in a Burma forest. The amber contained "micro-monsters" – fleas, biting midges, mosquitoes, and the like – that must have preyed upon the dinosaurs. Further, these micropredators surely would have also infected the dinosaurs, so seemingly invincible, with

---

10.1038/nature10231; C. M. Schlebusch et al., "Khoe-San Genomes Reveal Unique Variation and Confirm the Deepest Population Divergence in Homo Sapiens," *Molecular Biology and Evolution*, 37, no. 10 (2020), 2944–2954. DOI: 10.1093/molbev/msaa140. PMID: 32697301; PMCID: PMC7530619.

[30] Darwin to J. D. Hooker, July 13, 1856, Letter no. 1924, Darwin Correspondence Project, www.darwinproject.ac.uk/letter/DCP-LETT-1924.xml.

[31] Darwin to A. Gray, May 22, 1860, Letter no. 2814, Darwin Correspondence Project, www.darwinproject.ac.uk/letter/DCP-LETT-2814.xml.

infestations of gigantic nematode worms that ate their organs, slowly killing them.[32]

Irenaean Theodicy and Process Theodicy

In contrast to many versions of Augustinian theodicy, several other theodicies are more accepting of mainline science, including evolutionary science. For example, philosopher John Hick's "soul-making theodicy" was inspired by the writings of Irenaeus in the ancient Eastern Church, who envisioned the original humans, Adam and Eve, not as perfect but as innocent and immature, and therefore needing to learn and grow. Irenaeus holds that God's purpose is that human beings, who were originally created in God's "image" (Greek: eikón), will be brought into God's "likeness" (Greek: homoiósis).[33] In short, God desires to bring forth personal creatures with a depth of understanding and character that can have conscious fellowship with him.

In the spirit of Irenaeus, Hick notes that the first stage of God's creative process was easy for omnipotence – bringing the physical universe into existence and eventually guiding evolution to produce organic personal–rational life. As he explains, "man emerged out of the forms of organic life," leading to the second stage in which "the uncompelled responses and willing cooperation of human individuals ... in the world in which God has placed them" provide opportunity for the maturation of their souls.[34] Central to Hick's view is that the world is an environment that is necessary for "soul making" – a context in which temptation, hardship, pain, and suffering all occur and to which we must respond. From this perspective, the entrance of sin into human existence was due more to weakness and immaturity in a challenging environment than to conscious, radical disobedience to God.

Hick's theodicy, like most other theodicies, is anthropocentric and treats animal pain only secondarily. Yet the soul-making environment, for Hick, clearly includes animals, although he discounts their pains by invoking a neo-Cartesian understanding:

---

[32] J. R. Schneider, *Animal Suffering and the Darwinian Problem of Evil* (London: Cambridge University Press, 2020), 40–41.

[33] Irenaeus, *Against Heresies*, 5.6.1.

[34] J. Hick, *Evil and the God of Love* (New York: Macmillan, 1985), 255.

Not only is the animal's experience not shadowed by any anticipation of death or by any sense of its awesome finality; it is likewise simplified, in comparison with human consciousness, by a happy blindness to the dangers and pains that may lie between the present moment and this inevitable termination; and again by a similar oblivion to the past.[35]

Again, here we face the ambiguity of whether animals have inner states similar to those of humans coupled with certain philosophical and theological instincts that tend to discount their inner states.

Regardless of what might be roughly correct in Hick's assessment, on balance, science challenges the sweeping claim that animals (at least the higher animals) do not possess the type of consciousness that experiences pain in a fashion similar to human experience or that can unify past and present experience in fear of some future pain.

On balance, science challenges the neo-Cartesian leanings in Hick's theodicy. The evidence suggests that animals (at least the higher animals) have a conscious experience of pain similar to humans along with the ability to unify past and present experiences in fear of some future pain. Although science has not been able to discover all of the neural correlates of animal subjective states, we have already seen that it still supplies a number of empirical facts about animal behaviors – such as avoidance behaviors, grieving rituals, and self-recognition – which plausibly suggest recognizable inner states.[36] Furthermore, the persistent question of why an omnipotent being would be unable to make a much less painful environment for humans readily applies to the case of animals as well.[37] On this point, Hick argues for the status quo – that is, no reduction in the present amount and intensity of pain among humans – because soul-making requires it; one could infer that he likewise believes that animal pain in the world, which is the context for human spiritual growth, cannot be reduced as long as God's goal is soul-making.

Process theodicy has always attracted some religious thinkers because it offers an escape from some uncomfortable questions about why God cannot

---

[35] Hick, *Evil and the God of Love*, 312–314.

[36] N. Clayton, T. Bussey, and A. Dickinson, "Can Animals Recall the Past and Plan for the Future?" *Nature Reviews Neuroscience*, 4 (2003), 685–691. See also I. Bradshaw, "Not by Bread Alone: Symbolic Loss, Trauma, and Recovery in Elephant Communities," *Society and Animals*, 12 (2004), 145–147.

[37] J. J. Lynch, "Theodicy and Animals," *Between the Species*, 13, no. 2 (2002).

eliminate evil. Whereas Augustinian and Irenaean theodicies were based on a standard theistic view of omnipotence, process theodicy is based on a particular type of nonstandard theism (or quasi-theism) that denies the classical concept of omnipotence in order to explain why God cannot eliminate evil. Process theology is inspired by the work of Alfred North Whitehead, who sought to fuse the modern scientific vision of reality – which includes quantum mechanics, relativity theory, and biological evolution – with a certain vision of God and the world.[38] Theodicy in the process tradition specifically asserts that God seeks to lead the world from simplicity and triviality into greater complexity and significance, although this process inevitably involves discord and pain. For process theodicy, then, pain is necessary to achieve the greater good of important and intense experience for finite beings.

Process theologian John Haught actually makes evolution the key to theology generally and to theodicy specifically. In *God after Darwin: A Theology of Evolution*, he denies that God is a controlling sovereign intelligence that can guarantee all desired outcomes for the world. The process God, who "renounces any claim to domineering omnipotence," is more accurately conceived as "a vulnerable, defenseless, and humble deity" who seeks to persuade the entities of the world to go in his preferred direction.[39] Haught emphasizes that God voluntarily withdraws from creation in order to allow room for its own evolutionary creativity. In regard to the problem of suffering and evil – which involves contingency and randomness, suffering and tragedy – Haught states that God "participates in the world's struggle and pain" while seeking to lead the world into its ideal future.

Philosopher Gary Chartier applies process theodicy to animal pain by denying that God directly wills animal pain and instead arguing the familiar process line that God seeks to persuade or lure the evolving creation in his ideal direction but that creatures do not always cooperate. For example, God may try to guide Rowe's little fawn away from the forest fire, but from the standard process theology perspective, the freedom of indeterminism could result in the fawn not cooperating.[40] Nevertheless, this nondetermined process can still contribute to the improving world, as Chartier explains:

---

[38] See A. N. Whitehead, *Process and Reality*, corrected edn. (New York: Free Press, 1985).
[39] J. F. Haught, *God after Darwin: A Theology of Evolution* (New York: Routledge, 2018), 51, 54.
[40] M. Maller, "Animals and the Problem of Evil in Recent Theodicies," *Sophia*, 48 (2009), 299–317.

> The process thinker can opt for a consequentialist account of divine
> goodness ... thus she can maintain that acts of predation can contribute to
> evolutionary advance, that God wills at least some such acts to foster
> evolutionary advance, and that God's decision to do so is consistent with belief
> in God's goodness.[41]

Although this consequentialist process theodicy for animal pain is interesting, it raises serious questions. Whatever the value of evolutionary advance that "embodies God's creative intentions," is it sufficiently clear or unarguable that the value of higher species appearing is worth millions of years of animal pain – which seems like an excessively long time, filled with untold suffering – until higher species emerged? Darwin opined that the emergence of new species evolving from the suffering of trillions of lower species cannot be reconciled with a Creator of unbounded goodness.[42] Ultimately, a consequentialist approach in process theodicy entails the global principle that God aims for the optimum sum total of value in the world, which may involve some negatives, such as pain for animals as well as humans.

Biological Pain and Worldview Comparison

Judging from our brief review of major Christian theodicies, it would appear that insights into the problem of animal pain are in short supply. To date, perhaps the best strategy in theodicy that engages the evolutionary science on animal pain is to assert that providential wisdom upholds a lawlike natural order, which includes the biological order. Such a claim keeps a certain sort of theistic teleological argument alive by pointing out that evolutionary pain and suffering – winnowing unfit organisms from the planet for several billion years – was necessary to bringing about more advanced creatures, which are of great value. In any case, each theodicy includes, if only by implication, some explanation for animal suffering, which can be evaluated in light of scientific and philosophical considerations. These theodicies can actually be viewed as hypotheses – as worldview-level hypotheses – that can be compared in terms of their explanatory adequacy to all the evolutionary data.

---

[41] G. Chartier, *Religious Studies*, 42, no. 1 (2006), 21.
[42] Cited in Murray and Ross, "Neo-Cartesianism and the Problem of Animal Suffering," 169.

Of course, theistic theodicies are not the only world-level hypotheses regarding animal suffering that are available to thoughtful persons. Nontheistic and non-Christian world-level hypotheses also offer explanations of animal pain in nature. Philosopher of science Philip Kitcher denies that the idea of a benevolent Creator can be reconciled with carnage in nature, which he takes as a strong reason for atheism.[43] In *River out of Eden*, Richard Dawkins alludes to Darwin's mention of the larvae of the Ichneumonidae parasitoid wasp, which eats caterpillars from the inside:

> If Nature were kind, she would at least make the minor concession of anesthetizing caterpillars before they are eaten alive from within. But Nature is neither kind nor unkind .... It is easy to imagine a gene that, say, tranquilizes gazelles when they are about to suffer a killing bite. Would such a gene be favored by natural selection? Not unless the act of tranquilizing a gazelle improved that gene's chances of being propagated into future generations. It is hard to see why this should be so, and we may therefore guess that gazelles suffer horrible pain and fear when they are pursued to the death – as most of them eventually are. The total amount of suffering per year in the natural world is beyond all decent contemplation.[44]

He then recounts a human tragedy to emphasize the poverty of theodicy:

> Theologians worry away at the "problem of evil" and a related "problem of suffering." On the day I originally wrote this paragraph, the British newspapers all carried a terrible story about a bus full of children from a Roman Catholic school that crashed for no obvious reason, with wholesale loss of life. Not for the first time, clerics were in paroxysms over the theological question [of how an all-loving, all-powerful God can allow such horrible events]. The article went on to quote one priest's reply: "The simple answer is that we do not know why there should be a God who lets these awful things happen. If the universe was just electrons, there would be no problem of evil or suffering."[45]

Dawkins points out that the horror of the crash confirms to a Christian that we live in a world of real good and evil. Dawkins presses even further,

[43] P. Kitcher, "The Many-Sided Conflict between Science and Religion," in W. E. Mann (ed.), *The Blackwell Guide to the Philosophy of Religion* (Malden, MA: Blackwell, 2005), 268.
[44] R. Dawkins, *River Out of Eden* (New York: HarperCollins, 1996), 131–132.
[45] Dawkins, *River Out of Eden*, 132.

On the contrary, if the universe were just electrons and selfish genes, meaningless tragedies like the crashing of this bus are exactly what we should expect, along with equally meaningless *good* fortune. Such a universe would be neither evil nor good in intention. It would manifest no intentions of any kind. In a universe of blind physical forces and genetic replication, some people are going to get hurt, other people are going to get lucky, and you won't find any rhyme or reason in it, nor any justice. The universe we observe has precisely the properties we should expect if there is, at bottom, no design, no purpose, no evil and no good, nothing but blind, pitiless indifference.[46]

Comparing theistic and atheistic explanations of human and nonhuman suffering, Dawkins concludes that nontheistic explanation is more adequate.

In a more technical way, philosopher Paul Draper makes the same comparative judgment that the apparent indifference of the universe to pain in sentient creatures supplies strong evidence against theism. Offering his own rendition of the evidential argument from evil, Draper keys specifically on the apparently random distribution of pains and pleasures among human and nonhuman animals as to be explained by each of two rival hypotheses. He then asks which of the two rival hypotheses explains the distribution better. The "Hypothesis of Theism" (T) is a world-level hypothesis asserting that an omnipotent, omniscient, and morally perfect person created the universe. The other world-level hypothesis – which Draper calls the "Hypothesis of Indifference" (HI) – asserts the following:

> HI Neither the nature nor the condition of sentient beings on Earth is the result of benevolent or malevolent actions performed by nonhuman persons.

HI is inconsistent with standard theism but consistent with naturalism as well as other religious view involving indifferent deities or powers. However, understandably, HI is typically interpreted in favor of naturalism and atheism, as with Dawkins earlier. Of course, in modern Western culture, a naturalist philosophical framework is the default worldview home of atheism.

In order to adjudicate between the two competing hypotheses – T and HI – Draper appeals to the notion of "antecedent probability," that is, the idea that the probability of a specific proposition being true can be antecedently judged higher on one assumption than on another. For Draper, a proposition

[46] Dawkins, *River Out of Eden*, 132–133.

stating the random distribution of pain and pleasure in the world is more probable given HI than given T. After all, the distribution appears to be random from a moral perspective, as Dawkins indicated, and not according to merit or desert. Instead, the distribution of pain and pleasure, as Draper points out, seems much more correlated to the biological goals of organisms, survival and reproduction. Pain warns animals of danger and possible death so that they can survive and reproduce.

Assessing the two competing explanations, Draper claims that the distribution of pain and pleasure in the world is much more antecedently likely on HI than on T. Or, where P is antecedent probability and O is a statement referring to the observations earlier regarding the pattern of distribution of pain and pleasure, we get the following symbolic statement:

$$P(O/HI) \; > \; P(O/T).$$

For Draper, we would not antecedently expect the actual distribution of pain and pleasure the world contains if T were true, but we would expect it if HI were true. On the evidence specifically as described, then, it is more reasonable to believe HI than to believe T.

In assessing the strength of Draper's argument *against* T and *for* HI, a number of critical considerations surface. For example, Draper simply assumes what theism should imply regarding the distribution of pain and pleasure: that God would distribute pleasures and pains, rewards and punishments among humans, according to moral character and conduct; that for animals God would make many or all of them happy. As it turns out, as Draper sees it, "many humans and animals experience prolonged and intense suffering and a much greater number are far from happy."[47] However, intellectually sophisticated theists point out that these are debatable assumptions. Even the book of Job, which poignantly raises the question of why there is innocent human suffering, reflects the idea that some higher virtues – such as faith, courage, wisdom, and perseverance – require adversity of various sorts. Situated deeply within the Judeo-Christian tradition, the book of Job deconstructs the twin mistaken ideas that a person's physical

[47] P. Draper, "Pain, Pleasure, and the Evidence for Atheism," in M. L. Peterson et al. (eds.), *Philosophy of Religion: Selected Readings*, 5th edn. (New York: Oxford University Press, 2014), 418.

and biological circumstances are an indicator of his or her moral condition and that happiness cannot include suffering.

With regard to animal suffering specifically, several Christian thinkers who accept theistic evolution have proposed theodicies to address animal pain in the context of an evolutionary world. To the degree that any one of these proposals is plausible, it might serve to diminish Draper's high-probability assignment to HI and perhaps even to increase the probability of T. Theologian Christopher Southgate argues that God created a world in which good and harm are inseparably connected but that God in Jesus identifies with the suffering in that world and will ultimately redeem animals, as well as humans, that have suffered.[48] Theologian Bethany Sollereder rejects the Augustinian view that the doctrine of the Fall explains animal suffering in an evolutionary world and, like Southgate, emphasizes the need for the eschatological completion of creation through the resurrection and restoration of virtually all biological creatures that have ever existed.[49] Religion and science expert Victoria Campbell cites the brain-opioid theory of social attachment for how animals in social species mitigate suffering – that comforting behaviors stimulate the release of opioids in the brain that alleviate pain and distress. She claims that such phenomena serve both as evidence against overly broad Rowe-type claims about the great amount of suffering in nature and as evidence for how a loving, relational God would probably structure a living physical world.[50]

Draper acknowledges that his exclusive focus on the distribution of biological pain and pleasure as strong negative evidence against theism does not engage any potentially positive evidence for theism. Yet his approach is more measured than Dawkins's approach, which maintains that the universe "has precisely the properties we should expect if there is, at bottom, no design, no purpose, no evil and no good, nothing but blind, pitiless indifference." Yet, in philosophy as well as science, the ideal in assessing the explanatory power of a hypothesis is that we should consider the "total evidence" – that we make an all-things-considered judgment.

[48] C. Southgate, *The Groaning of Creation: God, Evolution, and the Problem of Evil* (Louisville, KY: Westminster John Knox Press, 2008).

[49] B. Sollereder, *God, Evolution and Animal Suffering: Theodicy without a Fall* (London: Routledge, 2019).

[50] V. Cambell, "A Philosophical, Scientific, and Theological Analysis of the Problem of Creaturely Suffering: Towards a New Perception of God and Pain," unpublished PhD Thesis Middlesex University, London School of Theology (April 2020).

So, on the other side, theists essentially claim that there appear to be a number of important properties of the universe that we would not antecedently expect if naturalistic atheism is true but would expect if theism is true. In pursuing the ideal of total evidence, these impressive phenomena that theism supposedly explains better must be taken into account in the assessment of theism and naturalism as worldviews. Theists, for instance, claim that theism is the best explanation for why there is something rather than nothing – that is, the fact that there is a universe at all – because they appeal to God's creative purposes whereas naturalistic atheists typically appeal to purposeless chance. Other important phenomena that theists also claim to be antecedently more probable on theism than on naturalistic atheism would be the lawlike character of nature as well as features of personhood such as consciousness, rationality, and our fundamental ability to make moral judgments. The theist's reasoning is that these things are more likely to occur in a reality created by a self-existent God who is orderly, conscious, rational, and moral than in a naturalistic reality that exists by pure chance and in which these fascinating properties of the universe arose through exceedingly improbable coincidence.

We cannot in this volume assess the comparative explanatory power of theism and naturalism relative to all evidence but rather invite the interested reader to review the debate between Michael Peterson and Michael Ruse, two philosophers of science, who address the total evidence from these two opposing worldview perspectives.[51] What we can reasonably say is that the specific problem of animal pain, while the evidence is significant, does not appear to be the sovereign deciding factor in the evaluation of the explanatory power of theism and naturalism.

---

[51] See the following paired articles: M. Ruse, "The Naturalist Challenge to Religion" and M. L. Peterson, "The Encounter between Naturalistic Atheism and Christian Theism," in Peterson et al. (eds.), *Philosophy of Religion*, 427–450. For a book-length treatment, see Peterson and Ruse, *Science, Evolution, and Religion*.

5   Progress, Purpose, and Providence

The history of the universe – from the Big Bang to *Homo sapiens* – is nothing short of breathtaking. The whole process started off with a violent explosion, went through various stages of cosmic structuring and cooling as galaxies and their solar systems developed, produced at least one planet – Earth – that contained life-supporting conditions and eventually produced life leading to humanity as we know it. In reflecting on the extended development of physical reality, important questions arise regarding whether the universe in general displays a discernible directionality and even whether biological reality is goal-oriented, perhaps with the aim of bringing forth *Homo sapiens*.

In this chapter, we sort through a cluster of issues surrounding the question of directionality. How would we describe the direction? What kinds of change or directionality count as progress in some way? Can progress be identified or measured? Although questions of evolutionary directionality and progress might seem to be empirical matters, extrascientific values and concepts inevitably play into some of these discussions and must be philosophically evaluated. Religion also has a major point of contact with science on these questions: theological concepts related to divine involvement and potential guidance toward some purpose must interact with current knowledge about how the physical world works.

Is There Directionality in the Universe?

A general description of the history of the universe – which includes the development of life – supplies the basis for addressing the question of directionality. We begin with a brief sketch of the major developmental stages of physical reality as science has revealed them. The cosmos began 13.75 billion years ago with the cataclysm of the Big Bang scattering hydrogen, helium, and lithium throughout the early universe. These elements eventually aggregated into dense giant stars, which became stellar cookeries that formed carbon, oxygen, nitrogen, iron, and other heavier elements. These in turn were later distributed everywhere when the giant stars expired and exploded, causing the chemical complexification of the universe. Cosmic structure developed further as comets and asteroids formed and stars and their planetary systems collected into galaxies. Although all materials at that point were inorganic, the universe was not static and went through identifiable changes, which raises the question of direction.

Forming about 4.5 billion years ago, Earth developed life-supporting conditions, allowing increasingly complex molecules to develop, utilize some kind of energy source, and result in the first self-replicating molecule, which we can date back to almost 4 billion years. From single-celled life, multicellular life eventually emerged about 2 billion years ago. The Ediacaran fauna flourished, setting the stage for what was to come. In what is known as the Cambrian explosion (or Cambrian radiation) just more than half a billion years ago, early representatives of most major animal phyla appeared in the fossil record. Over enormous spans of time, rich varieties of life forms appeared, becoming more complex and more taxonomically diverse. In the animal world, we get fish, amphibians, reptiles, dinosaurs (actually fancy reptiles), birds (actually fancy dinosaurs), mammals, monkeys, apes, and humans. And there is a similar story for the plant world. Through this process, increasingly sentient organisms appeared that were aware of their environment and could sense and respond.

To visualize the sweeping description of the major stages of development discussed earlier, while leaving pursuit of further detail to the reader, consider the sketch in Figure 5.1.

Although the chart in Figure 5.1 abbreviates an enormous amount of scientific information, the overall picture it paints is seen by many as suggesting directionality in the sense of movement over time toward greater complexity, an idea that begs for philosophical examination.

We cannot talk of the major aspects of the development of physical reality – cosmic, stellar, chemical, planetary, galactic, and so forth – without puzzlement at how it all began and developed. Each new stage is a significant transition in the history of the universe, a major threshold crossed. Science continues to advance our understanding of the processes involved at all

The Big Bang

[the universe studied by science begins]

⇩

Formation of Giant Stars and

Chemical Complexificationt of the Universe

[the material building blocks of the universe are developed]

⇩

Formation of Planets and Galaxies

[the large-scale structure of the universe is established]

⇩

Life on Earth

[the organic arises from the inorganic]

⇩

Sentient Life

[organisms appear that sense and respond to their environment]

⇩

Self-Conscious, Rational Life

[beings appear that have self-awareness and a variety of
intellectual capacities]

Figure 5.1 Development of the Universe.

levels – stellar nucleosynthesis, for example, is now well understood, whereas other processes, such as abiogenesis, are less so. The point is that science continues to answer a lot of questions regarding the unfolding history of the complex universe – and, in doing so, fills in the factual details answering the question, "How did we get here from there?" Philosophical disagreement arises, then, when we try to answer the metaphysical question,

"Why did we get here from there?" The metaphysical discussion involves teleology and brings worldview differences into sharp focus.

Were Humans Inevitable?

In scientific terms, the universe can produce life because it has already produced life, and, furthermore, it can produce humans because it has already produced humans. In the trajectory of physical reality, the most complex object in the universe – the human brain – was produced and is the hallmark by which we assign high status to humans, with all of their powers of rational thought, moral judgment, and social interaction. Questions surrounding whether the universe had to produce humans – whether humans were inevitable somehow – take us to the heart of the dispute between naturalism and theism. Contemporary naturalists explicitly affirm that the universe occurred by chance, that chance is ultimate and stands behind even the existence of physical laws and the ways the world unfolds. By contrast, theists assert that humans were purposed by God to be special in the order of creation.

We know the physical facts quite well. Paleontology and molecular biology have shown us very clearly the details of the human lineage.[1] Around 6 million years ago, the lineage leading to *Homo sapiens* diverged from the one leading to our closest living relatives, chimpanzees. About 2 to 3 million years ago, our genus, *Homo*, appeared in the fossil record. Our genus increasingly gained self-consciousness and intelligence – controlling fire and making stone tools. Some lineages began migrating out of Africa into Asia and Europe over the following hundreds of thousands of years. Around 350,000 years ago, anatomically modern *Homo sapiens* entered the archaeological record, though without the full suite of modern behaviors. By about 50,000 years ago, *Homo sapiens* had become behaviorally modern and were leaving Africa in large numbers, following earlier *Homo* species (such as Neanderthals and Denisovans). What is the best interpretation of all of these facts?

The question of interpretation is not merely a scientific question; it is also a philosophical question. Interestingly, early evolutionists such as Charles Darwin's grandfather, Erasmus Darwin, actually thought that humans were

[1] M. Ruse, *The Philosophy of Human Evolution* (Cambridge: Cambridge University Press, 2012).

inevitable. In a rather bad poem, he advanced the view that humans are the necessary aim of the universe, monad to man, as they used to say:

> Imperious man, who rules the bestial crowd,
> Of language, reason, and reflection proud,
> With brow erect who scorns this earthy sod,
> And styles himself the image of his God;
> Arose from rudiments of form and sense,
> An embryon point, or microscopic ens![2]

As a deist, Erasmus Darwin thought that God started the universe and then let it unfold through unbroken law.

Charles seemed to hold the same view, as the final words of the *Origin* eloquently illustrate:

> Thus, from the war of nature, from famine and death, the most exalted object which we are capable of conceiving, namely, the production of the higher animals, directly follows. There is grandeur in this view of life, with its several powers, having been originally breathed by the Creator into a few forms or into one; and that, whilst this planet has gone cycling on according to the fixed law of gravity, from so simple a beginning endless forms most beautiful and most wonderful have been, and are being evolved.[3]

Darwin himself suggested that complexity might just increase naturally and bring about ever-higher life forms, but he saw "no necessary tendency" to do this. Paradoxically, the evolutionary process that Darwin outlines in detail is riddled with contingencies, making it arguably incompatible with an inevitable rise of humankind. Nevertheless, many people mistakenly conflate Darwin's views with those of his contemporary Herbert Spencer, who taught that there is a universal law of progress; Spencer interpreted biological evolution as an inevitable upward climb as the "fittest" survive and "advance." Although Darwin objected to Spencer's "law" as unscientific, this interpretation became labeled "social Darwinism" and inspired ideas of "superior individuals," even the Nazi idea that a "superior race" could be created through eugenics.

In his mature thought, Darwin held that we cannot make qualitative judgments between species. In the margin of his copy of an early evolutionary tract by the Scottish publisher Robert Chambers (*Vestiges of the Natural History of*

---

[2] E. Darwin, *The Temple of Nature* (London: Johnson, 1803), 1.5, lines 309–314.
[3] C. Darwin, *On the Origin of Species*, 2nd ed. (London: John Murray, 1860), 490.

Creation), Darwin scribbled, "Never use the word higher & lower – use more complicated."[4] These sentiments are shared by countless evolutionists today, at both the biological and philosophical levels, for at least a couple of reasons. First, natural selection is not a drive or force but a de facto occurrence – the features that help an organism survive in one situation may not in another – and in this sense it is relativistic. As the famous paleontologist Jack Sepkoski emphasized, intelligence is only one possible adaptation among tetrapods: in a herd fleeing predators, running fast may be a better adaptation than greater intelligence.[5] Second, the raw building blocks of evolution – the genetic variations on which selection works – are random not in the sense of being uncaused but in the sense of not occurring according to need. There is no empirical evidence that these variations are directed toward the need for "higher" humans. In any case, divine guidance is not a subject for science but a metaphysical interpretation of the evidence provided by science, a matter we continue to explore.

Among evolutionary biologists who have no theistic commitment, Stephen Jay Gould stressed the central role of contingency in saying that the evolutionary tree of life, with its upward direction, should be replaced by the "evolutionary bush," which displays great divergences and diversity rather than a specific direction, all due to the chanciness of evolution. As Gould states,

> [A]bandon progress or complexification as a central principle and admit the strong possibility that H. sapiens is but a tiny, late-arising twig on life's enormously arborescent bush – a small bud that would almost surely not appear a second time if we could replant the bush from seed and let it grow again.[6]

Using a different analogy, Gould remarked that, if we rewound the tape of evolution and then played it forward again, the significant contingency of the evolutionary process would make it extremely unlikely that we would get humans.[7] Negative conclusions about progress to higher species,

---

[4] Cited in T. Shanahan, The Evolution of Darwinism: Selection, Adaptation, and Progress in Evolutionary Biology (Cambridge: Cambridge University Press, 2004), 288.

[5] M. Ruse, Monad to Man: The Concept of Progress in Evolutionary Biology (Cambridge, MA: Harvard University Press, 1996), 486.

[6] S. J. Gould, "The Evolution of Life on Earth," Scientific American, 271, no. 4 (1994), 91.

[7] S. J. Gould, Wonderful Life: The Burgess Shale and the Nature of History (1989; repr. New York: W. W. Norton, 2007), 14, 45ff.

let alone about necessary progress, are based on the tremendously important role of chance in evolutionary biology. Making the point bluntly, the French biologist Jacques Monod states that evolution proceeds by "pure chance, absolutely free but blind."[8]

Not all biologists who adopt a naturalist philosophical perspective are nondirectionalists. Many regard the overall sweep of evolution toward complexity as progress because they picture the struggle for survival as an "evolutionary arms race," in which every adaptation in one species (great speed, thick shell) is matched by an adaptation in another species (greater speed, stronger boring power). Richard Dawkins, for instance, endorses evolutionary progress for these reasons: "Directionalist common sense surely wins on the very long time scale: once there was only blue-green slime and now there are sharp-eyed metazoa."[9] In the long run, those beings with the biggest brains will end up on top. Gould eventually came to a similar view with his "drunkard's walk" model. His reasoning was that evolution is like a drunkard walking along a sidewalk, bounded on the one side by a brick wall and on the other side by the gutter; eventually the drunkard will end in the gutter, simply because he cannot go through the brick wall.[10] There is a boundary to the simplicity of life but no boundary to the complexity. In synthesizing their independent work, paleontologist Daniel McShea and philosopher Robert Brandon also argue for an upward momentum to life's history by introducing the very non-Darwinian "zero-force evolutionary law" – ZFEL for short. They state, "In any evolutionary system in which there is variation and heredity, in the absence of natural selection, other forces, and constraints acting on diversity or complexity, diversity and complexity will increase on average."[11]

The Cambridge paleobiologist Simon Conway Morris, who is famous for his groundbreaking work on the fossils of the Burgess Shale, is a Christian theist who believes scientifically in evolutionary directionality, which he speaks of as "convergence." He argues that there are many instances of

[8] J. Monod, *Chance and Necessity: An Essay on the Natural Philosophy of Modern Biology*, trans. A. Wainhouse (New York: Knopf, 1971), 112–113.

[9] R. Dawkins and K. Krebs, "Arms Races between and within Species," *Proceedings of the Royal Society of London, Series B*, 205 (1979), 508.

[10] S. J. Gould, *Full House: The Spread of Excellence from Plato to Darwin* (New York: Harmony Books, 1996), 149–151.

[11] D. McShea and R. Brandon, *Biology's First Law: The Tendency for Diversity and Complexity to Increase in Evolutionary Systems* (Chicago IL: University of Chicago Press, 2010), 3.

evolutionary convergence, meaning that organisms that are unrelated some-times take similar paths. For instance, in the early Pleistocene, we find unrelated species of the saber-toothed tiger – same body types but some are placental mammals and others marsupial. Conway Morris offers this as one among many pieces of evidence that there is, independent of organisms themselves, a niche for predators with the shearing teeth that the saber-tooth possesses. Morris thinks that it is simply a matter of an evolving line finding that particular niche and moving in. Perhaps, then, there is a niche for intelligence that was waiting to be found and was eventually found by our own species.[12] However, fuller discussions of whether there actually are objective niches waiting to be found, as well as why complexity per se must automatically be adaptive, cannot be pursued further here.

Scientific and Metascientific Questions

The standard scientific model of the development of physical reality provides a factual narrative of increasing complexity at all levels. On the biological level alone, evolution has led to organisms generally becoming larger in size, more complex in terms of number of cells and functions, more taxonomic-ally diverse, and more energetically intensive. Philosophically, however, there is much about the facts regarding increasing complexity to parse and interpret.

In the field of biology, for example, we have a number of unsettled questions. Does the notion of complexity pertain to a number of individual parts or functions, to isolated organisms, or to their interrelations in ecosys-tems? One dimension of these sorts of questions is of course scientific, but in another dimension such questions invoke extrascientific definitions, con-structs, and values that must be addressed philosophically. Among these metascientific questions are the following: Is complexity among biological organisms an indicator of anything else, such as progress? Does greater complexity in organisms constitute progress in the sense that it is "better"? Such questions can only be settled by appealing to various extrascientific commitments that are philosophical in nature – first, to designate the end point toward which something is supposedly moving, and, second, to supply

---

[12] S. C. Morris, *Life's Solution: Inevitable Humans in a Lonely Universe* (Cambridge: Cambridge University Press, 2003).

some criteria or metrics by which movement of an aspect of nature toward that end point may be adjudicated. Now, some of this may depend on biological construals of the organism. From the perspective of replication, evolution has typically not been viewed as progressive, because replication only perpetuates what random variation serves up.

However, with reference to the attributes reflecting organismic function – such as homeostatic precision, sensory acuity, metabolic scope, and so forth – there is a sustained tradition of viewing evolutionary history as progressive. Yet even here, there is room for further clarification: if progress means an increase in the mean (or biased replacement, in contrast to random diffusion away from a minimum wall), then it is not clear that even such characteristics as sensory acuity have changed progressively. Clearly, specifying the valued target and determining the metric for progress turn out to be complicated undertakings, intertwined with philosophical viewpoints and subject to anthropocentric and even political biases.

Closely related to the question of whether evolution is progressively directional is the question of whether evolution in some sense has a purpose. Obviously, we can have directionality without purpose, but can we have purpose without directionality? This consideration readily divides into two aspects. One aspect is whether the evolutionary process is purposive or teleological in the sense of entailing target orientation or intrinsic finality, which would be analogous to the developmental trajectory or homeostatic regulation of an organism. Even if one grants a target orientation to organisms, there is at present no theoretical or empirical warrant for this understanding of evolution.

Nonetheless, various religious and quasi-religious perspectives affirm intrinsic purpose in the living world, including, prominently, process theology and the Gaia hypothesis. Since we discussed the process theological vision of God seeking to lure the world into greater complexity and greater significance in Chapter 4, let us summarize the central commitments of the Gaia hypothesis, which was formulated in the 1970s by atmospheric chemist James Lovelock. Naming the hypothesis after Gaia, the primordial Earth goddess ("Mother Earth") in Greek mythology, Lovelock based it on conceiving of Earth itself as an integrated system. His essential postulate was that living organisms interact with their inorganic environment forming a synergistic, self-regulating, complex system that has the intrinsic purpose of maintaining and perpetuating the conditions for life on Earth. The Gaia hypothesis has been criticized by the scientific community as poorly estab-lished because it rests on selective examples chosen to produce the conveni-ent experimental results.[13] Nevertheless, the Gaia-type thinking about the holistic nature of Earth has been influential in, and has sometimes even been spiritualized by, the global ecology movement of the last several decades, a topic we discuss more fully in Chapter 10.

Another aspect of our consideration of purpose is not whether evolution itself pursues a purpose but whether it is fit for a purpose in the sense of extrinsic finality. In other words, are the characteristics of evolutionary history such that it could be reasonably construed as reflecting (though not demonstrating) God's ends or purposes? Although there is a complex empir-ical dimension to this question – pertaining to how the characteristics of evolutionary history are described – it is also deeply theological. One import-ant theological response in the early twentieth century was the evolutionary eschatology of Pierre Teilhard de Chardin, a French philosopher, paleontolo-gist, and Jesuit priest, who theorized that humanity is evolving, mentally and socially, toward an ultimate spiritual unity. The Omega Point, as he called it, is the final destination of the universe, required by both science and theology, as creation connects fully with God. Biological evolution – from single-celled organisms to animals, to animals with complex nervous systems, to *Homo sapiens* that developed intelligence – is caught up in the larger evolutionary trajectory of physical reality toward God.[14] Chardin's cosmological claim was that this ultimate fate of the universe is required by the fundamental laws of physics; his theological claim was that the Omega Point could be explained by the Christian concept of Christ as the *logos* of God, who calls himself in the book of Revelation "the Alpha and the Omega."[15] Before the Catholic Church adopted an officially positive view of science and evolution at the Vatican II Council in 1965, Pope Pius XII issued a warning that Chardin's writings did not comport well with trad-itional Catholic theological dogmas. Scientists, including biologists, also criticized Chardin for oversimplifying, if not sanitizing, the scientific data.

[13] J. W. Kirchner, "The Gaia Hypothesis: Conjectures and Refutations," *Climatic Change*, 58 (2003), 21–45.
[14] P. Teilhard de Chardin, *The Phenomenon of Man*, trans. B. Wall (New York: Harper Perennial Modern Thought, 2008), 29.
[15] Revelation 1:8; cf. Colossians 1:17.

Obviously, the discussion of directionality, progress, and purpose involves value commitments and metaphysical constructs as part of an interpretive framework for the scientific facts. In *The Descent of Man*, Darwin, always the faithful scientist, allows himself a moment of philosophical reflection on the evolutionary progress of the human species:

> Man may be excused for feeling some pride at having risen, though not through his own exertions, to the very summit of the organic scale; and the fact of his having thus risen, instead of having been aboriginally placed there, may give him hope for a still higher destiny in the distant future. But we are not here concerned with hopes or fears, only with the truth as far as our reason permits us to discover it; and I have given the evidence to the best of my ability. We must, however, acknowledge, as it seems to me, that man with all his noble qualities, with sympathy which feels for the most debased, with benevolence which extends not only to other men but to the humblest living creature, with his god-like intellect which has penetrated into the movements and constitution of the solar system – with all these exalted powers – Man still bears in his bodily frame the indelible stamp of his lowly origin.[16]

Darwin's contemplative statement is fascinating – recognizing the possibility that humanity's humble origin may still prepare the way for a nobler destiny in the distant future – and thus leaves open a purposive religious interpretation while neither affirming nor denying it as far as science can determine.

In opposition to positive interpretations of purpose in evolution, philosophical naturalists instead interpret the scientific facts as revealing that the universe and humanity itself have no purpose and are ultimately produced and driven by chance, thus making impossible progress toward any goal, particularly an extrinsic goal or end willed by deity. Bertrand Russell, influential atheistic philosopher in the first half of the twentieth century, was frank about his nihilistic perspective, maintaining that the world revealed by science is "purposeless, more void of meaning" than anything we can imagine:

> Man is the product of causes which had no prevision of the end they were achieving; ... his origin, his growth, his hopes and fears, his loves and his beliefs, are but the outcome of accidental collocations of atoms .... [All] the

---

[16] C. Darwin, *The Descent of Man*, vol. II (London: John Murray, 1871), 405.

noonday brightness of human genius [is] destined to extinction in the vast
death of the solar system, and . . . the whole temple of Man's achievement
must inevitably be buried beneath the debris of a universe in ruins.[17]

At rock bottom, the universe for Russell is "a meaningless dance of atoms."

Debates over Purpose and Providence

The topic of providence serves as a kind of overlay on the topics of progress
and purpose. While one key question is whether the facts of evolution may
be reasonably construed as concordant with God's purposes, a deeper question is whether God did *in fact* purposefully guide evolution. Wrestling with
these questions requires reflection on the nature of divine action, including
its relation to divine knowledge and power and to nature and natural laws.
As our discussion develops, we will consider on the religious side both
interventionist and noninterventionist theories of divine action in the world.
On the nonreligious side, we will consider views that deny both progress and
purpose and deny that any divine being is or could be involved.

In the philosophy of biology, the subject of divine action in the living
world was famously discussed in correspondence between Darwin and his
close American friend, the noted Harvard botanist Asa Gray. At the time, the
natural theology tradition in both England and America guaranteed that the
general public and many academics would perceive a tension between descent and design. For Gray, the resolution of the tension began when Darwin
was working intently on his theory and solicited statistical information from
Gray about the distribution of North American flora. During their
exchanges, Darwin confidentially shared the outline of his theory as early
as 1857, and the two continued to correspond. By the time the *Origin of Species*
appeared in 1859, Gray was already convinced the Darwin had made a
significant contribution to solving the "species problem," but he was also
convinced that the solution would be seen as a threat to Paley's design
argument and therefore to theism. Gray took it as his mission to argue in
various venues that Darwin's theory was compatible with theism – that
natural selection could be harmonized with the traditional design argument – by positing that God "guided" the evolutionary process.

[17] B. Russell, "A Free Man's Worship," in P. Edwards (ed.), *Why I Am Not a Christian and Other Essays on Religion and Related Subjects* (New York: A Touchstone Book, 1957), 107.

Darwin appreciated Gray's assistance in influencing the public not to think dichotomously in terms of conflict, but in private he objected to Gray that there was no evidence for design that benefits individual organisms. For Darwin, natural selection had replaced an "interfering God" who originates species and guides their adaptation. The debate that ensued is complex and nuanced, but a few salient points are worth mentioning. First, the exact conceptions of the design argument at play in nineteenth-century England were particularly conditioned by the cultural context. In Victorian England, Paley's rendition of the argument reflected the Newtonian and deistic view that the machinelike universe runs by unbroken laws decreed by God, which in turn produced species in fixed forms that do not change. Darwin's critique offered a purely natural interpretation of the origin of species that replaced design as Paley described it. The second point is that Darwin increasingly took the element of chance involved in selection to be incompatible with the divine design of organisms, writing to Gray,

> I am inclined to look at everything as resulting from designed laws, with the details, whether good or bad, left to the working out of what we may call chance. Not that this notion at all satisfies me. I feel most deeply that the whole subject is too profound for the human intellect.[18]

In another letter to Gray in 1860, Darwin endorsed "designed laws" but argued for "undesigned results" – partly because the apparent noninvolvement of God in the suffering and death of persons and animals suggested to him that God would not get involved in the morally less important task of constructing the minute details of each species.[19] In short, Darwin thought that chance and divine activity were mutually exclusive, whereas Gray thought they were not. To an extent, their points of view did not quite connect because Gray thought of design as providing a more general plan whereas Darwin thought of design as applying to specifics.[20]

---

[18] Darwin to A. Gray, May 22, 1860, Letter no. 2814, Darwin Correspondence Project, www.darwinproject.ac.uk/letter/DCP-LETT-2814.xml.

[19] Darwin to A. Gray, July 3, 1860, Letter no. 2855, Darwin Correspondence Project, www.darwinproject.ac.uk/letter/DCP-LETT-2855.xml.

[20] A. Gray, "Natural Selection Not Inconsistent with Natural Theology," in A. H. Dupree (ed.), *Darwiniana* (Cambridge, MA: The Belknap Press of Harvard University, 1963), 121–122; Darwin to A. Gray, November, 26, 1860, Letter no. 2998, Darwin Correspondence Project, www.darwinproject.ac.uk/letter/DCP-LETT-2998.xml.

Historically, there is not just one design argument but several, each reflecting a distinctive philosophical approach and cultural context. Gray thought of purposive providential activity in nature in noninterventionist terms, while Darwin was working against interventionist understandings of God's involvement. Contemporary proponents of Intelligent Design (ID) theory indicate that they accept common descent and natural selection for many organic structures but insist that we can identify certain organic structures that are "irreducibly complex" because they cannot have occurred by random variation and selection, as we saw in Chapter 3. Essentially, ID posits interventionist providential activity in the biological realm that brings about particular organic structures, which in effect is miraculous. Classically, a miracle is defined as the violation of the laws of nature by a divine being, which means that for ID the insertion of irreducibly complex structures in biological history would be empirical evidence of design. Some read the comment of Christian philosopher Alvin Plantinga as reviving Asa Gray's contention: that God somehow guides genetic variations (mutations), or at least those variations that will result in certain important evolutionary lines.[21]

In contrast to those theists who offer some sort of interventionist theory, other theists try to explain the providential guidance of evolution in non-interventionist terms. Thomas Tracy, Nancey Murphy, and Robert Russell are major theistic thinkers who propose noninterventionist special divine action.[22] Russell's particular proposal is that God guided mutations through the quantum level where the absence of strict deterministic laws allows room for God to operate – for God to "feed," in a sense, new events into the macro world. Philosopher of physics Bradley Monton states the general explanatory strategy:

[21] See A. Plantinga, *Where the Conflict Really Lies: Science, Religion, and Naturalism* (New York: Oxford University Press, 2011), 116, 359n14.

[22] N. Murphy, "Divine Action in the Natural Order: Buridan's Ass and Schrödinger's Cat," in R. J. Russell, N. Murphy, and A. R. Peacocke (eds.), *Chaos and Complexity: Scientific Perspectives on Divine Action* (Vatican City: Vatican Observatory Publications, and Berkeley: Center for Theology and the Natural Sciences, 1995), 325–356; T. Tracy, "Creation, Providence, and Quantum Chance," in R. J. Russell et al. (eds.), *Quantum Mechanics: Scientific Perspectives on Divine Action* (Vatican City: Vatican Observatory Publications, and Berkeley: Center for Theology and the Natural Sciences, 2002), 259–292; R. J. Russell, "Divine Action and Quantum Mechanics: A Fresh Assessment," in Russell et al. (eds.), *Quantum Mechanics*, 293–328.

So how could God act in the world without intervening? One way for God to do this is by acting at the indeterministic quantum level. For example, if there's some quantum process that has a 10 per cent chance of yielding outcome A, and a 90 per cent chance of yielding outcome B, God can, in a particular instance of this process, decide which outcome will result, without violating any laws.[23]

Interestingly, this fascinating proposal would make the influence of God real but empirically undetectable, which means that it remains a metaphysical theory without a strong scientific connection. One wonders if this sort of attempt presses too hard to wed Christian and theistic ideas to current science, just as the Catholic Church wedded itself to Ptolemaic science in the late Middle Ages. Further, just as Ptolemaic science was overturned, it is very possible that our understanding, say, of quantum physics could change. The fact that there are various interpretations of how quantum physics applies to reality – from deterministic to indeterministic – might serve as a caution not to intertwine any particular theory about the mechanics of natural operations with a theological position.

Seeking to transcend the interventionist/noninterventionist debate, the British Anglican theologian and biochemist Arthur Peacocke relies on the classical Christian understanding of God's immanence in creation – that God is present within the world while not ontologically identified with the world:

If the world is in any sense what God has created and that through which he acts and expresses his own inner being, then there is a sense in which God is never absent from his world and he is as much in his world as, say, Beethoven is in his Seventh Symphony during a performance of it.[24]

In making this point, he states that the immanence of God requires a stronger theological emphasis on the "evolutionary character of God's creative action," which he characterizes as moving from "the 'hot, big, bang' to humanity." For Peacocke and other theologians, the mainstream of Christian orthodoxy has always understood divine providence to include chance, seen as nondetermined contingency within the created world, such that chance in evolution, in interplay with lawlike regularity, raises no new problems for providence that were not known before the rise of the biosciences.

---

[23] B. Monton, "God Acts in the Quantum World," in Jonathan Kvanvig (ed.), *Oxford Studies in Philosophy of Religion*, vol. 5 (New York: Oxford University Press, 2014), 167.

[24] A. Peacocke, *God and the New Biology* (London: J. M. Dent & Sons, 1986), 96.

For scientific naturalists, however, there is still no metaphysical frame-work, ancient or modern, that could possibly offer a satisfactory concept of the divine relation to natural processes in general or evolutionary processes in particular. When evolutionary naturalists place great emphasis on radical chance – which shades into declarations of the ultimacy of chance – they rule out in principle any providential activity. George Gaylord Simpson reflects this thinking in his remark (Chapter 3) that humanity resulted from natural processes that have no intentional or purposed ends. Breaking with his own NOMA principle (Chapter 1), Gould could not help but opine that providence and radical evolutionary contingency are absolutely incompat-ible.[25] Clearly, the issue of evolution and divine purpose involves serious, complicated, and unresolved questions for religion. Equally clearly, issues on the science side, such as the types of chance involved in evolution, require further elucidation before conclusions with metaphysical and theological import are firmly drawn.[26]

Is Emergence a Viable View?

Talk of direction, progress, and even purpose in the natural world may be fruitfully connected to the philosophical discussion of emergence, particu-larly in a period when strict philosophical reductionist views are thought to involve major difficulties. Of course, as a philosophy of science, neither reductionism nor emergentism has an immediate effect on the daily busi-ness of doing science. Yet the philosophical debate between reductionism and emergentism has deep relevance for understanding the essence of natural science in terms of the explanations it seeks and the realities it studies. On the one hand, in mid-twentieth-century discussions, logical positivism strongly reinforced reductionism and thus inspired the "unity of science" movement for the natural sciences, which sought a single, unifying theory of all other theories, from general relativity and quantum theory to models of biological evolution. Philosopher W. V. O. Quine sup-ported this reductionist approach, once remarking that his ontology pre-ferred "desert landscapes" by designating as real only those empirical

---

[25] Gould, *Full House*, 18. See his comment that humans are a "momentary cosmic accident."
[26] See M. L. Peterson and M. Ruse, *Science and Religion: A Debate about Atheism and Theism* (New York: Oxford University Press, 2016), 111–112, 138–139. See these references for types of chance.

entities that ground all scientific work.[27] Along with most other physical-ists, he posited the basic particles recognized by physics as the most funda-mental constituents of reality.

On the other hand, in the past several decades, the general concept of emergence has become central in the ongoing conversation. The drift of the current discussion increasingly affirms the appearance of genuinely novel properties at certain levels of organizational complexity that were not pre-sent at lower, less complex levels, and it denies that these new properties are reducible ontologically to lower-level entities or reducible explanatorily to lower-level concepts. Such thinking has led to the concept of "superveni-ence" – and talk of these properties supervening upon certain complex systems. Emergence sees the universe as a hierarchy of levels that develops in dynamic ways as each level makes possible the appearance of the next, more complex level. Our earlier sketch of stages of development in the universe paves the way for more technical discussion of the levels of emer-gence. In fact, biophysicist Harold Morowitz has identified no fewer than 28 distinct levels of emergence in natural history, from the Big Bang to the present.[28] Interestingly, these commitments entail, among other things, a rejection of both dualism and mechanism. Lying in between vitalism at one extreme and reductionist determinism at the other, emergentism has become a lively topic in the philosophy of biology.

Two broad categories of emergence must be recognized – "strong emer-gence" (ontological) and "weak emergence" (epistemological). Strong emergentists maintain that genuinely novel causal agents or causal processes come into existence in the unfolding history of the physical world. Weak emergentists claim that, as new patterns emerge, the fundamental laws by which they operate are still ultimately physical, such that it is only at the level of description that we utilize emergent categories. The present discus-sion will focus on strong emergentism, accepting the point by philosopher of science Timothy O'Connor that "an emergent's causal influence is irredu-cible to that of the micro-properties on which it supervenes: it bears its influence in a direct, 'downward' fashion in contrast to the operation of a

---

[27] W. V. O. Quine, "Epistemology Naturalized," in *Ontological Relativity and Other Essays* (New York: Columbia University Press, 1969), 69–90.

[28] H. Morowitz, *The Emergence of Everything: How the World Became Complex* (New York: Oxford University Press, 2002).

simple structural macro-property, whose causal influence occurs via the activity of the micro-properties that constitute it."[29]

Although concepts of emergence can be traced through assorted past thinkers, such as Aristotle and Hegel, it was Cambridge philosopher C. D. Broad who articulated a strong antidualist theory of emergence early in the twentieth century. In his book *The Mind and Its Place in Nature*, Broad explained how to think about emergence:

> [W]e have to reconcile ourselves to much less unity in the external world and a much less intimate connection between the various sciences. At best the external world and the various sciences that deal with it will form a kind of hierarchy. We might, if we liked, keep the view that there is only one fundamental kind of stuff. But we should have to recognize aggregates of various orders.[30]

Traditionally, of course, "life," "consciousness," "mind," and "agency" have been considered good candidates for genuine emergents within the physical world.

Among early twentieth-century emergentists, Conway Lloyd Morgan was arguably the most influential in engaging specifically biological evolution. Unable to accept what many call Darwin's "continuity principle," which eliminates any "jumps" in nature, Morgan emphasized that emergence embraces the fact that evolution is "punctuated" by the occurrence of radically new life forms as stages in the evolutionary process. In this regard, Morgan resembled Alfred Russel Wallace who also disagreed with Darwin. However, unlike Wallace, Morgan argued in *Emergent Evolution* not for a "God of the gaps" but for a frank admission of the discontinuities revealed in the scientific evidence from natural history. For Morgan, these discontinuities are seen in the multiple levels that have occurred, with new wholes or objects or agents in their own right.[31]

The current state of the discussion is not settled, of course, with some voices denying strong emergence and its most prominent feature of downward causation. The construction of scientific explanations of such phenomena as chemical bonding by quantum mechanics and the rise of molecular

---

[29] T. O'Connor, "Emergent Properties," *American Philosophical Quarterly*, 31 (1994), 97–98.

[30] C. D. Broad, *Mind and It's Place in Nature* (London: Kegan Paul, Trench, Trubner & Co., 1925), 77.

[31] C. L. Morgan, *Emergent Evolution* (London: Williams and Norgate, 1931), 207.

biology have greatly encouraged reductionists, who claim that putatively emergent phenomena are in principle reducible to interactions among the lower-level constituents. Philosopher John Searle, for example, envisions the existence of "causally emergent system features" – for instance, liquidity, transparency, and consciousness – which are properties of a system exercising downward causality that cannot be predicted from facts about its parts but, nonetheless, can be explained purely by the causal relations of the parts.[32] Based on Searle's approach, then, consciousness is a fascinating case of an emergent state while still being explicable in terms of the underlying behavior and interaction of neurons.

In the continuing debate over emergentism and reductionism, a number of thinkers consider the phenomenon of emergence to be highly relevant to understanding religion in an age of science. For example, Niels Gregerson, a scholar in science and religion, interprets emergence to be consistent with monotheism (Judaism, Christianity, and Islam), which sees God as causally prior to the universe while working within it. Yet Gregerson acknowledges that emergentist interpretations fall across a wide spectrum. At one extreme, we place "flat religious naturalism," which expresses awe at nature but denies any supernatural being, a view somewhat reminiscent of the nineteenth-century American transcendentalists such as Ralph Waldo Emerson and Walt Whitman. In this vein, biologist Ursula Goodenough reflects the following sentiment in *The Sacred Depths of Nature*:

> Emergence. Something more from nothing but. Life from nonlife. Like wine from water, it has long been considered a miracle wrought by gods or God. Now it is seen to be the near-inevitable consequence of our thermal and chemical circumstances.[33]

Another type of naturalistic view that assumes that nature is prior to god may be labeled "evolving theistic naturalism." The opinion of early twentieth-century philosopher Samuel Alexander was that God actually emerges from natural processes due to the striving of space–time to achieve deity.[34]

---

[32] J. Searle, *The Rediscovery of Mind* (Cambridge, MA: MIT Press, 1992), 111.
[33] U. Goodenough, *The Sacred Depths of Nature* (New York: Oxford University Press, 1998), 28–29.
[34] S. Alexander, *Space, Time, and Deity*, 2 vols. (London: Macmillan, 1920).

One of the stronger interpretations of emergence on the more traditional theistic side has been developed by German theologian Wolfhart Pannenberg in his "theology of the future." In arguing for what could be labeled "eschatological theism," Pannenberg develops his idea in the following way:

> Every event throws new light on earlier occurrences; this now appears in new connections. This fact seems to have possessed considerable weight for the thinking of the Israelites. Their thinking implied, one might say, an eschatological ontology: if only the future will teach us what is the significance of an event, then the "essence" of an event or occurrence is never completely finished in the present. Only after the larger connection of occurrences to which an event belongs has been completed can the true essence of the individual event be recognized. In the last analysis, only the ultimate future will decide about its peculiarity.[35]

A prime example of inclusion in an emergent reality in which the whole system retrospectively defines the parts more completely would be the brain's inclusion of circuits that allow the capacities of individual neurons to be realized.

Theists of various stripes have offered interpretations of emergence by revising theology in a scientifically informed way. For classical Thomists, God is working through secondary causes in the evolution of all things; for process theists, God is seeking to persuade nondetermined entities to emerge according to his ideal aims. In all such interpretations of emergence, God is not static and neither is his creation – in fact, God is the initiator or guide of a process in which nature is endowed with powers to produce new levels of reality over time, new types of objects with new types of causal powers, which require new types of explanation at each unique level. As philosopher Philip Clayton and physicist Paul Davies have explained, it might have turned out that systems are simple aggregates or at least act in ways completely determined by their parts, but the renewed talk of emergence in our day signals that reductionist/determinist thinking cannot make total sense of what science is telling us and seems to open a path for theological reflection.[36]

---

[35] W. Pannenberg, *Toward a Theology of Nature: Essays on Science and Faith*, Ted Peters (ed.) (Louisville, KY: John Knox Press, 1993), 83.

[36] P. Clayton and P. Davies (eds.), *The Re-Emergence of Emergence: The Emergentist Hypothesis from Science to Religion* (New York: Oxford University Press, 2006), ix–33.

The Concept of Information in Biology

One corollary of the idea of emergence is that somehow information increases over time as new levels of entity arise. The concept of information in biology, which has been widely employed in biology since the 1960s, is particularly interesting in relation to theistic belief. It is concerned not only with how information is encoded and transmitted but also with how information increases and complexifies over time, thus providing fruitful lines of exploration for evolutionary biology, particularly molecular and developmental biology.[37] After all, despite the obvious evolutionary advantage of simple life-forms (such as rapid replication), a wide range of complexity levels have been achieved by living systems, giving them potential to cope with challenges in a selective environment.[38]

As evolutionary biologists John Maynard Smith and Eörs Szathmáry have suggested, there is a "progression" in how information is packaged and replicated across the major evolutionary transitions.[39] Gould represents a common view in maintaining that complexity is not a trait that evolution explicitly favors, raising the question of why bacteria, the simplest life form, are not the only life form. As noted earlier, a complexity wall exists just below bacteria because life-forms below this wall fail to subsist, which means that random fluctuation is more likely to produce more complex forms, although complexity is not explicitly favored.[40]

However, since the 1950s, information theory, as developed by American mathematician and engineer Claude Shannon, has shown itself to be relevant and important to many subfields of biology – from studies of perception and cognition to more basic aspects of biological theory, such as explanations of hormones and other cellular products as regulatory "signals" and descriptions of the causal role of genes in metabolic processes in terms of "transcription" and "translation."[41] Biologists routinely speak of the "developmental program"

[37] J. M. Smith, "The Concept of Information in Biology," *Philosophy of Science*, 67, no. 2 (2000), 177–194.

[38] Gould, *Full House*.

[39] J. M. Smith and E. Szathmáry, *The Major Transitions in Evolution* (New York: Oxford University Press, 2010). See also E. Szathmáry and J. M. Smith, "From Replicators to Reproducers: The First Major Transitions Leading to Life," *Journal of Theoretical Biology*, 187, no. 4 (1997), 555–571.

[40] J. T. Bonner, *The Evolution of Complexity by Natural Selection* (Princeton, NJ: Princeton University Press, 1988).

[41] C. E. Shannon, "A Mathematical Theory of Communication," *Bell System Technical Journal*, 27, no. 3 (1948), 379–423.

that unfolds when organisms progress from zygote to adult. It is remarkable that contemporary biology as an overtly materialist science has come to include deeply intentional or semantic concepts like this in its work. This has at times raised philosophical tensions among biologists and, as we will see, also plays a role in challenges to mainstream biology advanced by the ID movement.

In large part, the language of information theory became commonplace in biology for historical and practical reasons. The work to characterize the means by which DNA functioned as a hereditary molecule, and the subsequent elucidation of the so-called genetic code by which sequences of DNA form functional proteins, took place during a time when information theory and increasing computation were ascendant. Since molecular biology and computer science each came into their own alongside each other, it was natural for biologists to borrow the language of information theory to describe what they were discovering. Not least in this time was a desire to establish biology as a theoretical and quantitative science, rather than merely a descriptive one. Information theorist and physicist Hubert Yockey, a pioneer in applying information theory to biology in the 1950s and 1960s, exemplifies this desire:

> Although there are many fields of biology that are essentially descriptive, with the application of information theory, theoretical biology can now take its place with theoretical physics without apology. Thus biology has become a quantitative and computational science . . . . By employing information theory, comparisons between the genetics of organisms can now be made quantitatively with the same accuracy that is typical of astronomy, physics, and chemistry.[42]

The broader question is whether the language of information theory as it pertains to biology is merely a metaphor or is an accurate description of biological reality. While Yockey argued strongly for the latter,[43] many biologists disagree. To explore this question, a brief excursus into the details of

---

[42] H. P. Yockey, *Information Theory, Evolution, and the Origin of Life* (Cambridge: Cambridge University Press, 2005), ix.

[43] On this point Yockey is emphatic: "Information, transcription, translation, code, redundancy, synonymous, messenger, editing, and proofreading are all appropriate terms in biology. They take their meaning from information theory (Shannon, 1948) and are not synonyms, metaphors, or analogies." Yockey, *Information Theory, Evolution, and the Origin of Life*, 6.

how DNA works will be necessary, and, tellingly, the reader will notice how difficult it is to describe these workings without resorting to analogy and metaphor.

A famous milestone in the molecular biology revolution was solving the structure of DNA as a double helix, where two complementary strands wind around one another and either can act as a template for the other. A DNA double helix is made up of two long molecules, which have a sequence of four chemical components commonly called nucleotide "bases." Most non-biologists know them only by their letter abbreviations: A, C, G, and T – but these are in fact organic molecules in their own right that are strung together in a long chain to make one of the strands of a double helix. What Watson and Crick famously solved was the structure of the helix – showing that, as the two strands wind around one another, chemical affinities pair up A and T, and similar affinities pair up C and G.[44] These affinities are what allow a single strand to act as a chemical template for constructing a second strand – giving a physical, chemical mechanism for duplicating DNA.

Having solved this problem, biologists immediately saw the next one: how this "information" – a specific sequence of DNA bases – could be used to determine the sequence of a protein. While DNA is made up of only 4 repeating units (bases), proteins are made up of 20 (amino acids). While DNA is excellent at replication with high fidelity, it does not have the ability to take on the numerous shapes and functions that proteins have. So, biologists reasoned, the sequence information from DNA must somehow be converted to the sequence of proteins. Since DNA and proteins use different "languages," the term "translation" became the metaphor for this process.[45] Since DNA has only four bases, it was appreciated early on that some combination of DNA bases would be needed to specify an amino acid – and the resulting system came to be described as a "code" and approached much like a cryptography problem.

After much work, the "genetic code" was "cracked" and shown to be common to all living things. Sets of three DNA bases, called "codons," were used to specify the 20 amino acids, as well as other signals important for the

[44] J. D. Watson and F. H. C. Crick, "Molecular Structure of Nucleic Acids: A Structure for Deoxyribose Nucleic Acid," *Nature*, 171 (1953), 737–738.

[45] An early example of the translation metaphor can be seen in G. Gamow, "Possible Relation between Deoxyribonucleic Acid and Protein Structures," *Nature*, 173 (1954), 318.

translation process. Further study showed that the translation process worked through an intermediate molecule connecting a given amino acid to a given codon. This connection seemed to be arbitrary – that in principle, any codon could have been assigned to any amino acid – further giving the impression that DNA was indeed a code much like what a human would invent. This particular angle has been heavily exploited by the ID movement to argue that biological information is indeed "information" in this sense – and thus the product of a designer – rather than a metaphor or analogy for complex chemical interactions. This line of argumentation is an old one within this movement, stretching back to its overtly creationist antecedents.

Under consideration are the ultimate origins of the DNA/protein transla-tion system as well as the specific DNA sequences found in various (separ-ately created) living organisms. An explicitly creationist, antievolutionary work published in 1984 – just as the ID movement was taking shape and moving to a form to evade constitutional challenge in the United States – is representative of this view. In attempting to make a case for only limited speciation from originally created prototypes, the authors state the following:

> However, in terms of the mechanism of limited variation, the application of information theory to the genetic machinery should prove the most promising. The crucial factor will be delineation of the necessity of *intelligent* design in the structuring of the informational content and grammar of each prototype. This will indicate the necessity not only of intelligence in originating the genetic code in the broad universal sense but also, in the specific sense, of the unique adaptive programs of each prototype.[46]

These arguments continue to be major ones within the ID movement and have changed little since they were first articulated.[47]

Molecular biology, however, has greatly advanced our understanding on both topics since the 1980s. The origin of the genetic code is wrapped up with the origin of life, as the genetic code is universal to all living organisms. The

---

[46] Lester and Bohlin, *The Natural Limits to Biological Change*, 167; emphasis original. We note that Bohlin later became an ID research fellow.

[47] See S. C. Meyer, *Signature in the Cell: DNA and the Evidence for Intelligent Design* (New York: HarperOne, 2009); D. R. Venema, "Seeking a Signature: Essay Book Review of Signature in the Cell," *Perspectives on Science and Christian Faith*, 62, no. 4 (2010), 276–283; S. C. Meyer, *Darwin's Doubt: The Explosive Origin of Animal Life and the Case for Intelligent Design* (New York: HarperOne, 2013).

origin of life, as we have discussed, is a frontier of science and not fully understood – and thus the origin of the code is similarly shrouded in mystery at present. Despite this, biologists continue to make progress. One primary finding of the last decade is that the genetic code is not arbitrary, as those in the ID movement claim: it too has chemical affinities within it, though exactly how those affinities were involved in the formation of the code remains an area of inquiry. What has been discovered is that there are chemical/physical associations between some codons (or in some cases, the "mirror image" of codons) and the amino acids they code for. In a chemically arbitrary code, there would be no expectation to find such a pattern. These findings are now well accepted, but debate about how they relate to the ultimate origin of the code remains, as might be expected for a research frontier. However, the idea that the genetic code is chemically arbitrary, and thus a code like what a human would design, has not found support. What we call the genetic code has evidence within it that it arose through chemistry, and as such, the consensus view of biologists is that calling it "information" is an analogy if one intends "information" in the humanly generated sense.

Similarly, the creationist claim that the "information content" of various species reflects their special creation as separate designs has not found support. As geneticists have sequenced and compared the genomes of many organisms, clear signals of shared ancestry are abundant in nature, including for humans.[48] This work has also showed us how information in DNA can be added to or subtracted from a genome over time. For example, the lineage leading to modern-day vertebrates experienced large gains in information through duplication of entire chromosome sets on two distinct occasions, giving a large number of gene copies that could acquire new functions through mutation and selection. Conversely, information can be lost: primates like ourselves, for example, have lost many of the genes devoted to the sense of smell (though we are compensated somewhat by new information giving us tricolor vision). Biologists thus think of a genome (or species) as a population-level phenomenon changing over time: both gaining information and losing information, much like a language.[49] As such, it is more proper to

---

[48] D. R. Venema and S. McKnight, *Adam and the Genome: Reading Scripture After Genetic Science* (Grand Rapids: Brazos, 2017), 29–42.
[49] Venema and McKnight, *Adam and the Genome*, 22–24.

think of a species, and its DNA information, as a *process* rather than a static entity.[50]

The process-like nature of species and the contingency-laden process by which information is added to or removed from a genome through mutations might lead one to conclude that Gould was ultimately correct in his estimation that rerunning the tape of life would produce entirely different results[51] and perhaps lend support to a dysteleological understanding of evolution. Though evolution is widely understood by nonbiologists to be dominated by chance events – and thus to be highly contingent – there is much evidence to support the idea that evolution is also greatly constrained and repeatable, or convergent. Examples of convergence abound: in Australia, there are marsupial "equivalents" of familiar placental mammals such as mice, moles, wolves, and even flying squirrels. All marsupials, however, are more closely related to each other than to any placental mammal. Despite chance-driven information gains and losses that were independent in these separate lineages, natural selection shaped each "pair" in a similar way to a similar niche. As such, evolution can be seen as very much nonrandom in overall effect – a point that might be useful for a theistic, teleological view. These ideas have been explored and developed in depth by Simon Conway Morris, whose book *Life's Solution* remains the key work in this area.[52] Evolution, and the increasing complexity of information it has produced over time, results from a mix of chance and constraint – and as such lends itself to both teleological and dysteleological interpretations among biologists. Such interpretations, however, go beyond what science can establish – that evolution is both contingent and convergent – into metaphysical considerations of how these observations relate to nontheistic and theistic interpretations.

Though many biologists find the information metaphor helpful – and indeed, it is very challenging to talk about complex biology without resorting to it – some biologists contend that ultimately these metaphors are harmful in that they cloud scientific thinking and provide an opportunity for pseudoscience:

> The simple empirical observation that science talk depends heavily on
> analogical thinking mandates that we examine the consequences of deploying

[50] See, for example, O. Rieppel, "Species as a Process," *Acta Biotheoretica*, 57 (2009), 33–49.
[51] Gould, *Wonderful Life*, 48.   [52] Morris, *Life's Solution*.

certain metaphors on how students and the public at large end up understanding science, and that we be attentive to the way pseudoscientists often seize upon these metaphors to foster misunderstandings.[53]

Of course, the understanding of information in biology as ultimately sourced in a physical–chemical basis and increased through natural means is perceived as a threat only by versions of theism (such as creationism and ID) that embrace a dichotomy between divine action and natural processes.

Thinking of God as found in the gaps not filled by science can lead to rejecting well-established science in order to find some auspicious "gaps" where God can be alleged to act. Though common and popular, this mode of thinking has also long been criticized from within theistic communities. The great German theologian Dietrich Bonhoeffer, writing from a Nazi prison, offered a cogent rejection of God-of-the-gaps thinking, and called for a more robust interaction between science and faith:

> Weizsäcker's book *The World View of Physics* is still keeping me very busy. It has again brought home to me quite clearly how wrong it is to use God as a stop-gap for the incompleteness of our knowledge. If in fact the frontiers of knowledge are being pushed further and further back (and that is bound to be the case), then God is being pushed back with them, and is therefore continually in retreat. We are to find God in what we know, not in what we don't know; God wants us to realize his presence, not in unsolved problems but in those that are solved.[54]

If continued reflection on the relation of God and biology is to be productive, it will require tandem commitments to the facts as biology discovers them to be and to the conceptual resources of the theological traditions involved.

---

[53] M. Pigliucci and M. Boudry, "Why Machine-Information Metaphors Are Bad for Science and Science Education," *Science & Education*, 20 (2011), 454.

[54] Bonhoeffer to Eberhard Betheg, May 29, 1944 in *Dietrich Bonhoeffer: Letters and Papers from Prison* (Minneapolis, MN: Fortress Press, 2010), 405–406.

Part II

Religion and Human Biology

6   Human Nature and Human Uniqueness

Homo sapiens is the only species that asks questions about its own existence, which means that we are self-interpreting animals. We humans have always been on a quest to know our own nature and to ask whether being human makes us significantly unique, qualitatively different from all other things, living and nonliving. Traditional philosophical traditions maintain that humans are special, and indeed, the highpoint of the natural world, while major religious traditions generally affirm human specialness based on our relation to the divine. In early modernity, however, the Galileo affair was seen as consolidating the "Copernican humiliation" of humanity; it initiated a great change in the way science understands the status of human beings, a change that has profoundly impacted philosophical and religious understandings.

Both science and religion, then, contribute to the ongoing quest for insight into our own humanity. As scientific information about humans advances and religious perspectives attempt to engage that information, a number of difficult questions arise. By what criteria would we judge human specialness? How does knowledge of our biology affect our view of our total humanity? How did the development of language influence the human evolutionary trajectory? Is rationality what sets humans apart? Can humanity's perceived difference in kind be properly interpreted to be a difference in degree from other animals? In the treatment to follow, we explore these and other questions.

Self, Soul, or Mind as the Human Essence

From ancient times to the present, humanity itself has, rather unsurprisingly, always been an important subject for reflection and study. One of the most influential theories about the essence of humanity comes to us from ancient Greece in Plato's idea that the core of our humanity – indeed, of the human person – *is* the rational soul, which is immaterial, both distinct and separable from the material body. In the *Apology*, Plato's soul–body dualism was expressed by Socrates as he tells the Athenian court – which had sentenced him to death for refusal to honor the polytheistic gods of Athens – that they could kill his body but could not touch his soul.[1] Aristotle, Plato's famous student, thought more holistically that a human being is a rational animal, an amalgam of rationality and animality, two elements that are conceptually distinguishable but not separable in reality. In developing his holistic view, Aristotle explicitly pointed out the key differences between types of souls – the nutritive soul (shared by all living things), the sensible soul (shared by all living things that can perceive their environment), and the rational soul (belonging only to humanity, which is capable of intellectual thought) – noting that each higher level of soul shares something with the lower. Throughout intellectual history, both humanists and religious believers have found in Plato's and Aristotle's views important ways to affirm humanity's special status on metaphysical grounds. Both Platonic and Aristotelian views of the human person found their way into Christian thinking in the Middle Ages – Plato's view through St. Augustine, Aristotle's through St. Thomas Aquinas – and both views today have representatives in the religious community.

Not all ancient thinkers, however, agreed that the soul is special and distinct from matter. The Greek atomist Democritus did not ascribe an intangible element to humanity, famously teaching that all reality is constituted by physically indivisible atoms in a void. Democritus and the other early atomists thought that the human soul was corporeal, composed of very subtle atoms but still destined like all corporeal things for disintegration. A materialist view of the human person was also affirmed in Epicurean philosophy in the third century BC and later advocated by Lucretius in his epic work *On the Nature of Things* in the first century BC.

At the other extreme from materialist views of the human essence, the Eastern religious traditions generally developed spiritualized or immaterialist understandings of humanity. According to Hinduism, for example, matter is unreal, the realm of *maya* (illusion) in which all humans are

---

[1] Plato, *Apology*, 41c–42a.

trapped until they find release from *karma*, which binds us to the cycle of birth–death–rebirth known as *samsara*. Furthermore, the human soul (*Atman*), considered as an individual person, is not ultimately real either but is rather an expression of the great Soul of the Universe (*Brahman*). As the *Upanishads* declare, "All creatures lose their separateness when they merge at last into pure Being. There is nothing that does not come from him. Of everything he is the inmost Self. He is the truth; he is the Self supreme."[2] So the human soul is spiritual and not material, but it is ultimately part of a great monistic or pantheistic reality. A noteworthy attempt to put Hinduism and science in dialogue on this exact topic is found in *Hindu Theology and Biology* by religion scholar Jonathan Edelmann, who offers a thorough exploration of the Hindu understanding of immaterial consciousness in relation to the standard materialist Darwinian understanding of consciousness.[3]

Historically, science–religion interactions in the West have typically run along very different lines. At the dawn of science in the early seventeenth century, Descartes projected a view of the world that continues today as a backdrop for all discussions of humanity. Descartes held that human beings are composed of two substances – a material substance extended in space (*res extensa*) and a thinking substance, an immaterial mind (*res cogitans*), which is the true core of the human person. The Cartesian outlook set the terms for debating how such radically different substances as mind and matter could interact, with Descartes himself suggesting that they interact at the pineal gland at the base of the brain. Occasionalists, who are also dualists, denied interaction in order to preserve the integrity of science in its study of the material world, contending that God coordinates mental states and brain states without their interaction.

Much of modern philosophy has sought an alternative to Cartesian dualism, often along the lines of materialism or physicalism. As mechanism in the sciences gained increasing acceptance, dualism was put under increasing pressure to explain what a human being is and what makes a human special. In the nineteenth century, the mechanist dictum was that all events in the world are "closed under physics," with no outside influence. Mechanists in the

---

[2] *Chandogya Upanishad* 6.10.1–3.
[3] J. Edelmann, *Hindu Theology and Biology: The Bhagavata Purana and Contemporary Theory* (New York: Oxford University Press, 2012).

nineteenth century, such as Thomas Henry Huxley, advocated epiphenomen-alism, which maintains that the conscious mind is a product of the physical system – the brain – although mental events can never influence the physical system. In the twentieth century, physicalist and behaviorist positions spoke of mind as a mere "ghost in a machine" and the idea of dualism as "Descartes's myth." Nevertheless, some philosophers found all forms of phys-icalism to be counterintuitive and thus implausible; religious thinkers found physicalism to be threatening to any semblance of human specialness. As we trace these issues here, our discussion will interact with the progress of science in explaining more facts about consciousness and the added pressure it places on locating the uniqueness of humanity in a soul or mind.

What about Origins?

Traditional religious understandings of humanity tend to associate the dis-tinctiveness of humanity not just with a self or soul but with the idea that the self or soul was created (or endowed) by God and somehow reflects God. Thus, the religious employment of some sort of dualism was generally very import-ant to setting humanity apart from the rest of the universe. As science developed and more phenomena came under the mechanistic model, the belief that we humans have any special, unimpeachable feature that guaran-tees uniqueness was shaken. Materialism seemed more in accord with science, and any dualism that posited some element beyond physical explanation became harder to maintain.

With the publication of Darwin's *Origin of Species* in 1859, the stage was set to ask the question "What is humanity?" from the perspective of natural history. Essentially, the origin of humanity was addressed scientifically rather than religiously – and the information and insights brought to bear were considerable. Evolutionist George Gaylord Simpson comments on how radical this new perspective was:

> The question "What is man?" is probably the most profound that can be asked by man. It has always been central to any system of philosophy or theology .... The point I want to make now is that all attempts to answer that question before 1859 are worthless and that we will be better off if we ignore them completely.[4]

[4] G. G. Simpson, "The Biological Nature of Man," *Science*, 152, no. 3721 (1966), 472.

Simpson's tacit scientism, which automatically affirms science as the only kind of knowledge and dismisses religious and philosophical efforts to understand our humanity as worthless, can be debated by those holding other perspectives. But Simpson's more basic point stands: it is undeniable that efforts since Darwin cannot succeed unless they take with great seriousness the place of our own humanity in the biological world. Of course, with the appearance of *The Descent of Man* in 1871, Darwin officially ushered in a new approach to the question of our humanity by making public that he thought the *Homo sapiens* was also a product of evolution.

In reaction to the scientific claim that humans were descended from a common ancestor along with chimpanzees and bonobos, Christian groups that took the Genesis creation story literally insisted that the biological account of origins would destroy human specialness. Biblical literalists believe that creation in general involves God's direct activity in bringing about creatures in their various fixed forms, but they regard the idea of humanity being specially created by God to be an extremely important event that grounds human specialness. For literalists, since Genesis teaches that "God created man in his own image,"[5] the theory of humanity's descent from lower animals would remove its special status in God's order.

The different reactions to the biological fact of human common descent make for interesting case studies in the epistemology of belief. On the side of biblical literalism, which is central to Protestant fundamentalism, the special creation of Adam and Eve was a nonnegotiable core belief. Predictably, various strategies were enacted in order to attack the idea of human evolution – among them, questioning the scientific methodology of evolutionists, questioning the fossil evidence for human animal ancestry, and linking evolution to atheism and secularism. Regarding methodology, for instance, the assumption of an ancient Earth by evolutionary theory was challenged by "young Earth" positions that embraced Bishop Ussher's biblical chronology supporting creation 6,000 years ago. The carbon-14 dating method supporting a very old Earth with very old fossils was under constant attack as flawed. For decades, the "missing link" debate was waged by fundamentalists claiming that the fossil record (particularly humanoid skulls) was insufficient to support human evolution from primates. Another fundamentalist attack on evolution was (and still is) to link it to

---

[5] Genesis 1:27.

perceived negative social outcomes, such as decline in religious faith and the rise of agnosticism, atheism, and secular humanism. After all, fundamentalists reasoned, if people believe that humans are merely descended from monkeys, why should they think that we are designed in the image of God?

On the nonreligious side, Victorian intellectuals wanting to free science from religion saw Darwinian evolution as a major ally because it offered an alternative account of human origins apart from religious teachings. Several other intellectual currents in the nineteenth century also contributed to the break with conventional, literal Christianity and helped solidify a formidable cultural resistance to religion. Huxley coined the word "agnosticism," arguing that claims about spiritual matters are not hard science and cannot possibly be known to be true. German scholars working on textual criticism of biblical documents confirmed that various passages were not meant to be literally true. Philosopher and anthropologist Ludwig Feuerbach, who strongly influenced the transition from conventional religion to positivism and atheism, declared that all religion, including Christianity, is a projection of our own ideals onto a god who does not exist in reality.

Occupying the middle ground were thinkers who had impressive credentials as scientists and accepted a broad Anglican perspective on Christian orthodox theology. Often accused of "accommodationism," some became authors of the Bridgewater treatises (see Chapter 3), which attempted to reconcile the geological record with Christianity and even parlay that evidence into natural theological arguments for God. Professor of medicine John Kidd, renowned philosopher of science William Whewell, and Oxford geologist and mineralogist William Buckland are all on the list of eight authors.[6] In this approach we have yet another variation in the epistemic dynamics caused by reacting to the challenge of Darwinian evolution to a long-held core belief.

Although the broad scientific picture of human origins has long been established, some debates about the facts of human origin continue today. But these debates are typically fueled by religious stereotypes of the Genesis story and of what is necessary to protect human dignity as special to God. Transcending such contentious debates over the facts of evolution, a number of religious responses today accept the science and refocus the debate on differences in philosophical interpretations of the facts – particularly between Christian and naturalist worldviews. Michael Peterson and

---

[6] *The Bridgewater Treatises I–VIII* (London: Pickering, 1833–1837).

Michael Ruse take this approach in *Science, Evolution, and Religion: A Debate about Atheism and Theism*, stipulating the facts of evolution and making the cases for their respective worldview perspectives.[7]

Wisdom counsels that accepting evolution as science is essential to any serious philosophical worldview. After all, from Darwin's day to ours, empirical confirmation of the theory of evolution has come from a myriad of facts in a wide variety of disciplines such as paleontology, comparative anatomy, biogeography, biochemistry, embryology, and ecology. Francisco Ayala, a world-famous biologist and Catholic believer, has stated, "Gaps of knowledge in the evolutionary history of living organisms no longer exist."[8] The universal tree of life – showing the pattern of continuity of all living organisms from a common ancestor to the present – is confirmed by hundreds of scientific studies published over many decades. Since the sequencing of the human genome was completed in 2003 – thanks to computer-assisted molecular biology – we can now trace common ancestry using the hereditary molecule, DNA.[9] The "language" of the genome involves the same 4 nucleotides and the same 20 amino acids in all living things, making common origin overwhelmingly probable. In fact, the degree of correspondence between the sequences of nucleotides in the DNA not only supports common ancestry but also allows us to reconstruct the phylogenetic tree of life. In these reconstructions, less closely related species exhibit less similarity in their DNA than more closely related species because more time has elapsed since their last common ancestor. Unsurprisingly, *Homo sapiens* is the most recent species, as shown by its DNA, which differs only by a small percentage from that of chimpanzees and bonobos. Human DNA even exhibits copying errors at the same points as theirs, again, making common ancestry the only logical conclusion.

Human Nature

Different thinkers and groups tie the question of human origins to the question of human nature. For those groups operating on a warfare model –

---

[7] M. L. Peterson and M. Ruse, *Science, Evolution, and Religion: A Debate about Atheism and Theism* (New York: Oxford University Press, 2016).

[8] F. J. Ayala, *Darwin's Gift to Science and Religion* (Washington, DC: Joseph Henry Press, 2007), 79.

[9] I. Ebersberger et al., "Mapping Human Genetic Ancestry," *Molecular Biology and Evolution*, 24, no. 10 (2007), 2266–2276.

fundamentalists and secularists alike – the claim that humans are descended from lower animals threatens the elevated status of humanity. Protestant fundamentalists cannot give up the argument that humanity was specially created without giving up their identity as a movement and thus must remain at odds with evolutionary science. Similarly, although accepting of evolution in general, Muslim Imam Shaykh Yasir Qadhi argues that Islamic theology requires that Allah miraculously inserted Adam into the natural order, replete with soul, conscience, and free will.[10] As the Qur'an states, "We created man from sounding clay, from mud moulded into shape."[11] It also affirms the existence of an original human pair, from which all generations of human being have come to inhabit the Earth: "Oh humankind! We created you from a single pair of a male and a female, and made you into nations and tribes."[12] Asserting the literal creation of two distinct original human beings might solve the problem of human specialness, if it were not scientifically completely untenable. For religious thinkers and groups who accept the facts of human evolution, the dualistic route is open for claiming that the human soul or spirit was endowed upon our biological form by God.

The Catholic Church, for example, signaled openness to evolution when the encyclical *Humani Generis* was issued in 1950:

> The Teaching Authority of the Church does not forbid that, in conformity with the present state of human sciences and sacred theology, research and discussions, on the part of men experienced in both fields, take place with regard to the doctrine of evolution, in as far as it inquires into the origin of the human body as coming from pre-existent and living matter – for the Catholic faith obliges us to hold that souls are immediately created by God.[13]

Later, in 1996, Pope John Paul II addressed the Pontifical Academy of Sciences, declaring that "we will all be able to profit from the fruitfulness of a trustful dialogue between the Church and science." More specifically, he stated that evolution is "more than a hypothesis" and "an essential subject

---

[10] N. Khan and Y. Qadhi, "Human Origins: Theological Conclusions and Empirical Limitations," Yaqeen Institute for Islamic Research, August 31, 2018, https://yaqeeninstitute.org/nazir-khan/human-origins-theological-conclusions-and-empirical-limitations/.

[11] Qur'an 15:26.   [12] Qur'an 49:13.

[13] Pius XII, encyclical *Humani Generis* (1950). Available online at www.vatican.va.

which deeply interests the Church."[14] Whatever the process that produced the human body turns out to be, however, John Paul II issued a warning against philosophical interpretations of human nature that are "materialist," "reductionist," or even "epiphenomenalist," for the Church holds that humanity is "created in the image and likeness of God." Thus, John Paul II reaffirmed with Pius XII the essential Catholic teaching that God "infuses" souls into humans – a miraculous event undetectable by science.

Interestingly, the famous Christian writer and thinker C. S. Lewis held a similar dualistic view, essentially maintaining that God somehow guided evolution until a creature with a sufficiently complex brain and neural structure emerged that could support a rational soul bestowed by God:

> For long centuries God perfected the animal form which was to become the vehicle of humanity and the image of Himself. He gave it hands whose thumb could be applied to each of the fingers, and jaws and teeth and throat capable of articulation, and a brain sufficiently complex to execute all the material motions whereby rational thought is incarnated.[15]

Lewis admits that the developing animal may have gotten progressively more capable in achieving purely material and animal ends but emphasizes that it was still just an animal. However, what made this animal form become human was God's endowment of a soul with the capacity to think about matters far above the mundane and immediate:

> Then, in the fullness of time, God caused to descend upon this organism, both on its psychology and physiology, a new kind of consciousness which could say "I" and "me," which could look upon itself as an object, which knew God, which could make judgments of truth, beauty, and goodness, and which was so far above time that it could perceive time flowing past.[16]

In his own way, Lewis agrees with John Paul II that only the soul is "able to ground the dignity of the person."[17]

Although scientific critics such as Stephen Jay Gould took this position as a great advance over creationist literalism, Richard Dawkins severely criticized it. In an essay entitled "You Can't Have It Both Ways," Dawkins labeled

---

[14] John Paul II, "Truth Cannot Contradict Truth: Address of Pope John Paul II to the Pontifical Academy of Sciences," *L'Osservatore Romano*, English edn. (October 22, 1996).

[15] C. S. Lewis, *The Problem of Pain* (1940; repr. New York: HarperCollins e-books, 2014), 73.

[16] Lewis, *Problem of Pain*, 73.   [17] John Paul II, "Truth Cannot Contradict Truth."

the "ensoulment" theory "casuistical double-talk," particularly objecting to the reaffirmation in *Humani Generis* that the mind is not an evolutionary product. Dawkins first summarizes the Church's position: "In plain language, there came a moment in the evolution of hominids when God intervened and injected a human soul into a previously animal lineage." Then, with disdain, he feigns curiosity regarding when God might have entered the evolutionary situation: "When? A million years ago? Two million years ago? Between *Homo erectus* and *Homo sapiens*? Between 'archaic' *Homo sapiens* and *H. sapiens sapiens*?" Dawkins stakes out his own position that there is no "great gulf between *Homo sapiens* and the rest of the animal kingdom" and that Pope John Paul II's view is "an anti-evolutionary intrusion into the domain of science."[18] Here Dawkins denies John Paul II's claim that "with man we find ourselves in the presence of an ontological difference, an ontological leap" – an "ontological discontinuity" that runs counter to the "physical continuity" studied by evolution.[19]

Dawkins continues the warfare image of science and religion and implicitly endorses a basic materialism about the human person. Analytic philosophers have put forward various more technical versions of materialism, as we shall see. While agreeing in their appeal to science for support of the view that the human person is entirely a material object, these versions differ on exactly how to articulate the point. Many materialists endorse identity theory, which holds that mental properties are identical to, and thus reducible to, internal bodily properties. Functionalism, on the other hand, takes mental properties to be identical to the causal roles played by these physical properties as they interact with the environment. Finally, eliminativism holds that the whole idea of inner mental states should actually be abandoned. Philosopher of neuroscience and materialist Patricia Churchland takes this view, maintaining that the traditional view of mental states such as beliefs, desires, and intentions is "folk psychology," which should be replaced in favor of neurobiological states.[20] Additionally, there are nonreductive and noneliminativist versions of materialism

---

[18] R. Dawkins, "You Can't Have It Both Ways," in P. Kurtz (ed.), *Science and Religion: Are They Compatible* (Amherst, NY: Prometheus Books, 2003), 208.

[19] John Paul II, "Truth Cannot Contradict Truth."

[20] P. Churchland, *Neurophilosophy: Toward a Unified Science of the Mind/Brain* (Cambridge, MA: MIT Press, 1986).

that postulate mental properties as distinct from, but supervening upon, physical properties.

Reductionist and materialist interpretations of the human person are generally associated with an atheistic position, but some religious thinkers have incorporated a material concept of the person into their theological perspective. For example, Christian philosopher of science Nancey Murphy has endorsed what she calls "nonreductive physicalism" based largely on neuroscientific discoveries suggesting that capacities previously attributed to mind, soul, or spirit as an immaterial element are now yielding to physical examination: "[N]euroscience has in a sense *completed* the Darwinian revolution, bringing not only the human body but the human mind as well, into the sphere of scientific investigation."[21] In essence, Murphy's nonreductive physicalism maintains that cognitive functions "emerge" from adaptive evolutionary processes. Clearly, the discussion of human nature impinges on the question of human uniqueness. If any version of materialism or physicalism is a correct interpretation of the data from neuroscience, then what becomes of the traditional distinctively human features, such as rational thought, free will, and awareness of the divine? Since Murphy is aware that the data from neurobiology can be seen as suggesting that we are all hardwired neural machines, she argues for a sophisticated theory of "top down processing" and "higher order evaluative systems" that change during maturation according to feedback from experience and society, displaying a remarkably "plastic" brain in the prefrontal cortex.[22] Numerous critics, religious and secular, charge that this sort of project is both bad theology and bad science, but we will postpone further investigation of the subject until later in this chapter.

Searching for Human Uniqueness

We now take up the pressing question of human uniqueness – particularly as it involves discussion of cognitive capacities that are supposedly distinctive.

[21] N. Murphy, "Human Nature: Historical, Scientific, and Religious Issues," in W. S. Brown, N. Murphy, and H. N. Malony (eds.), *Whatever Happened to the Soul? Science and Theological Portraits of Human Nature* (Minneapolis, MN: Fortress Press, 1998), 1.

[22] N. Murphy and W. S. Brown, *Did My Neurons Make Me Do It? Philosophical and Neurobiological Perspectives on Moral Responsibility and Free Will* (New York: Oxford University Press, 2007).

Using John Paul II's language, the issue in part boils down to the question of the "continuity" or "discontinuity" of human beings with the rest of nature. Most naturalists/materialists ally themselves with science in support of the continuity thesis: that human beings differ only in degree from other animals. By contrast, traditional philosophical and theological perspectives embrace the discontinuity thesis: that human beings differ in kind from all other things in nature. As we shall see, scientific evidence can be deployed for either view, and even for a middle view that grants both continuity and discontinuity in certain respects.

Darwin expressed an early preference for continuity in *The Descent of Man*, arguing for "no fundamental difference between humans and the higher mammals in their mental facilities" and identifying similarities in emotion, attention, and memory.[23] Since animals possess the same faculties in lesser degrees, Darwin reasoned that humans differ from other animals *only* in degree, albeit an impressive degree. Though he speaks even of our "god-like intellect," he also remarks that it still bears the indelible mark of our animal origins.[24] Philosopher of science Michael Ruse, like Dawkins, approves of Darwin's pronouncement that we should never call any creature "higher" – including humans – because evolutionary science provides no absolute scale of value to support such a judgment. Besides, scientists have documented similarities in anatomical structures, internal organs, DNA sequences, and many other features, showing that humans are closely related materially and structurally to the great apes. Some primatologists have even argued that language, which was traditionally thought to be distinctively human, can be learned to some extent by apes, and that basic tool use occurs among various primates, who, for example, employ rocks to crack nuts.[25] And we know that chimps kiss, laugh, and engage in group politics and goal-directed behaviors, as we explore more fully in Chapter 7.

Heralding what he sees as the triumph of the continuity thesis, well-known biologist Jerry Coyne stated that science throughout its history has delivered devastating blows to the theistic worldview but that Darwin delivered the most devastating blow of all:

---

[23] C. Darwin, *The Descent of Man*, 2nd ed. (London: John Murray, 1882), 66.
[24] Darwin, *Descent of Man*, 619.
[25] University of Birmingham, "Chimpanzees Can Learn How to Use Tools without Observing Others," *ScienceDaily*, September 28, 2017, www.sciencedaily.com/releases/2017/09/170928094204.htm.

[I]n 1859, when Charles Darwin published *On the Origin of Species*, [he] demolish [ed], in 545 pages of closely reasoned prose, the comforting notion that we are unique among all species – the supreme object of God's creation, and the only creature whose early travails could be cashed in for a comfortable afterlife.[26]

Coyne continues, "[L]ike all species, we are the result of a purely natural and material process." From the continuity thesis, Coyne concludes the mediocrity principle: that there is nothing special about human beings. Critics of this line of reasoning indicate that it is driven by tacit naturalist assumptions that select apparent physical continuities at all levels while ignoring various noteworthy cognitive and behavioral phenomena in humankind that show great discontinuity.

On the side of discontinuity, traditional philosophical and theological arguments have often been deployed to support the dualistic claim that, to be ontologically special, humans must possess a soul or mind or spirit distinct from the body. As John Paul II reflected theologically, this claim is compatible with the recognition of scientific facts of human evolution. In contemporary analytic philosophy, there is renewed discussion of mental phenomena, reviving hope for some type of dualism. One line of discussion that some had hoped would support dualism pertains to the "easy problem of consciousness," which involves how we discriminate and categorize stimuli, focus attention, integrate information, and the like, but these matters seem to be gradually yielding to brain science. Another discussion revolves around what philosopher of mind David Chalmers calls the "hard problem of consciousness," which is the profound difficulty of explaining the existence of phenomenal experiences (or *qualia*), such as the subjectively felt quality of emotion, mental images, and bodily sensations.[27] Thinkers inclined toward dualism find support in the inability of science to describe and explain subjective consciousness in objective third-person terms. Because of this difficulty, Colin McGinn, a contemporary philosopher of mind, has expressed skepticism that we can ever solve the hard problem of consciousness – that on this matter we have "cognitive closure" – because evolution did not equip our minds to understand such things.[28]

---

[26] J. Coyne, "Creationism for Liberals," *The New Republic* (August 2009), 34.
[27] D. Chalmers, "Facing Up to the Problem of Consciousness," *Journal of Consciousness Studies*, 2, no. 3 (1995), 201.
[28] C. McGinn, "Can We Solve the Mind-Body Problem?," *Mind*, 98, no. 391 (1989), 350.

In recent decades, philosophers who reject materialism, traditional dualism, and skepticism have proposed "emergent dualism" regarding the nature of consciousness and mind. On the one hand, they contend that substance dualism tends to support uniqueness by diminishing the importance of human animal history and biological function, which preempts a holistic account of human nature. On the other hand, they argue that materialism argues against uniqueness by discounting the elusive nature of many mental phenomena, thus effectively settling for reductionism in principle even though there are no convincing scientific reductive explanations for these phenomena. Christian philosophers William Hasker and Timothy O'Connor, for example, have independently argued for emergent dualism, defining "emergent" as a process where mental properties manifest themselves when the appropriate material constituents are configured in special, highly complex relationships.[29] These emergent properties are not observable in simpler configurations nor are they derivable from the laws that describe the properties of matter in simpler configurations. To use a rough analogy, a magnetic field and its properties are generated when the physical constituents of a magnet are in certain relationships – a sufficient number of iron atoms aligned. But the magnetic field is a real, existing, concrete entity, with new causal powers of its own.

In Chapter 2, we discussed emergentism as an interpretation of the nature of life, positioned between vitalism and materialism, but here that emergentism is a serious explanation of mental phenomena such as "consciousness" and "thought." Proponents of emergent dualism argue that it makes good sense of scientific findings as well as our common human experience of ourselves. The emergent dualist can agree with the materialist that potentially mental phenomena reside in the nature of matter but still avoid the mistake of denying what philosopher John Searle calls the "obvious facts about the mental," which are subjective states inaccessible to investigation.[30] Yet the emergent dualist also agrees with the substance dualist that the mind is distinct from the physical brain while avoiding the view that the mind is a nonmaterial entity, separable from the brain, with no systemic or internal relationship to the physical.

---

[29] W. Hasker, *The Emergent Self* (Ithaca, NY: Cornell University Press, 1999); T. O'Connor, "Emergent Properties," *American Philosophical Quarterly*, 31, no. 2 (1994), 91–104.

[30] J. Searle, *The Rediscovery of the Mind* (Cambridge, MA: MIT Press, 1992).

On this approach, it is not consciousness per se but particularly human rational consciousness that is viewed as a unique emergent reality that characterizes human beings. For emergentists, then, the human mind, while undoubtedly rooted in a distinctive brain size and structure that differ only in degree from our primate kin, also displays cognitive capabilities far beyond the cognitive abilities of primates. The list of distinctively human traits includes the ability for abstract thought (say, building systems of higher mathematics), for highly complex spoken language, and for contemplating the distant future, including alternative possible outcomes.

In academic theology and ordinary life, we apply the term "person" to the total psychosomatic unity that is a human being. As Arthur Peacocke, a biochemist and Anglican priest, says,

> There is therefore a strong case for designating the highest level – the whole, in that unique system that is the human-brain-in-the-human-body-in-social-relations – as that of the "person." Persons are *inter alia* causal agents with respect to their own bodies and to the surrounding world (including other persons).[31]

In evolutionary history, the emergence of the personal, with the salient characteristics of self-consciousness, rational thought, and agency, proves to be the paragon of what the various sciences see in emergence: whole–part influence. Various types of theists, including Hasker and O'Connor, argue that the emergence of this kind of personal being makes good sense in a universe created by a supremely rational personal agent. Supplementing basic theism, many Christian thinkers claim that the classic Christian doctrine of creation is readily interpreted to imply that God's plan was to bring forth holistic embodied personal agents. Furthermore, the classic Christian doctrine of the Incarnation affirms that God in the Second Person of the Trinity became bonded with a particular historical person, Jesus of first-century Nazareth. Conjoined with emergent dualism, all of this would imply that God bonded, intimately and forever, with a member of a species, produced by evolution, that had developed rational thought and moral agency.

---

[31] A. Peacocke, "Complexity, Emergence, and Divine Creativity," in N. H. Gregersen (ed.), *From Complexity to Life: On the Emergence of Life and Meaning* (New York: Oxford University Press, 2003), 197.

Contemporary Science and Our Humanity

Michelangelo's famous fresco *The Creation of Adam* on the ceiling of the Sistine Chapel depicts a bearded God – surrounded by an anatomically accurate drawing of the human brain – imparting through his forefinger the spark of life to the first human. The pictorial symbolism in this Renaissance painting links brain to mind, which is associated with the impartation of the divine image. Brain is part of the physical world, developed by the evolutionary process we share with higher animals; mind gives us capacities far beyond what any other creature displays. In a way, the painting revisits our question of whether humans differ in kind or degree from other animals, and it invites the judicious answer that they differ in both ways – that there are continuities and discontinuities. Interestingly, classical Christian theology actually entails that this should be the case: God made humans out of the material world, differing in degree from other material things, but God endowed them with higher capacities, differing in kind.

When it comes to science, the empirical evidence has frequently been interpreted philosophically to support the strong position of total continuity, but a more nuanced assessment of the evidence suggests that science itself gives a mixed answer as well – that there are both continuities and discontinuities between humans and the other animals. Let us develop a brief sketch of the scientific facts in this context. Consider first that an evolutionary perspective sees all organic change in general as the outcome of gradually unfolding natural history and sees the emergence of consciousness as being gradual as well. However, although the emergence of consciousness in the animal kingdom was indeed gradual, when humans came on the evolutionary scene, the rate of change increased dramatically, producing large, complex brains in a relatively short time, leading to remarkable development of mental ability. A striking fact that science teaches us is that the human brain is the most complex physical object in the entire universe. Compared to early hominids and hominins, a dramatic increase in cranial capacity allowed *Homo sapiens* to house larger brains, with a noteworthy ratio of brain-to-body mass, and much more specialized differentiation, such as the disproportionately large cerebral cortex, which is associated with reasoning and abstract thought.[32]

---

[32] S. Dorus et al., "Accelerated Evolution of Nervous System Genes in the Origin of *Homo sapiens*," *Cell*, 119 (2004), 1027–1040.

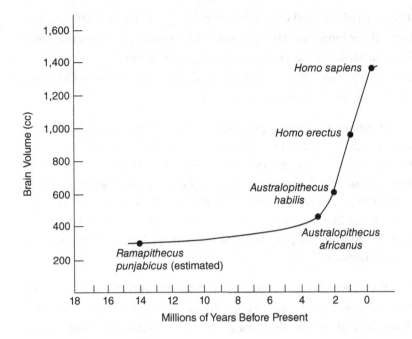

Figure 6.1 Growth Trajectory of the Human Brain

E. O. Wilson has stated simply as a matter of science that "the growth in intelligence that accompanied this enlargement was so great that it cannot yet be measured in any meaningful way."[33] For perspective, consider Figure 6.1 on brain size acceleration adapted from *The Ascent of Man* by paleoanthropologist David Pilbeam[34]:

In Figure 6.1, the growth trajectory of the human brain makes a virtual right-angle turn upward about 3 million years ago, leading Wilson to comment that "no organ in the history of life has grown faster."[35] Increased brain size and complexity is also linked to other important physical changes – for example, the lower placement of the larynx, which allows a much wider range of sounds for a vastly more complex language – characteristic of humans but arguably impossible to account for in terms of selective fitness. Of course, features such as brain size and complexity, and even nuanced sound

[33] E. O. Wilson, *Sociobiology*, 25th anniversary ed. (Cambridge, MA: Harvard University Press, 2000), 548.

[34] D. Pilbeam, *The Ascent of Man: An Introduction to Human Evolution* (New York: Macmillan, 1972).

[35] E. O. Wilson, *On Human Nature*, 25th anniversary ed. (Cambridge, MA: Harvard University Press, 2004), 87.

capability, can all be argued to be differences in degree. But the differences in degree are all the more interesting in light of important correlates that seem so different in kind, such as complex language and culture.

Reinforcing human specialness, brain research scientist Bruce Lahn concludes that humans are importantly unique:

> Human evolution is . . . a privileged process because it involves a large number of mutations in a large number of genes . . . . To accomplish so much in so little evolutionary time – a few million years – requires a selection process that is perhaps categorically different from the typical processes of acquiring new biological traits . . . . It required a level of selection that is unprecedented. Our study offers the first genetic evidence that humans occupy a unique position in the tree of life. Simply put, evolution has been working very hard to produce us humans.[36]

Although there are physical continuities between humans and all other animals, some would still argue for great discontinuity – say, between chimps using stones to crack nuts while humans cut diamonds and make fine jewelry.

Anthropologists relate greatly increased brain size with the "cultural explosion" (or creativity explosion, or cognitive explosion) of the Upper Paleolithic Period – the astounding result of the coevolution of brain, language, and culture. Connecting brain, language, and culture is an inevitably interdisciplinary activity as we continue to pursue the question of human distinctiveness. Many psychologists, linguists, and philosophers link anthropic specialness to the use of language and symbolic thought. Ian Tattersall of the American Museum of Natural History argues that increasingly complex language caused human thought to become much more efficient and powerful, allowing the transmission of ideas between minds and laying the groundwork for increasingly sophisticated cultures. The ability to communicate and to store information enabled accumulated learning and thus allowed even more growth of human rational powers and their applications to the world.[37] Hence, cultural evolution took humans along paths completely inexplicable in terms of purely biological evolution.

---

[36] B. Lahn interview by C. Gianaro, "Human Cognitive Abilities Resulted from Intense Evolutionary Selection, Says Lahn," *University of Chicago Chronicle*, 24, no. 7 (2005), http://chronicle.uchicago.edu/050106/lahn.shtml.

[37] I. Tattersall, "Innovation in Human Evolution," in J. Miller (ed.), *The Epic of Evolution: Science and Religion in Dialogue* (Upper Saddle River, NJ: Pearson Prentice Hall, 2004).

Around 45,000 years ago, abstract and symbolic thought notably accelerated – as evidenced, for example, in the spectacular cave paintings in France
and Spain from the Upper Paleolithic.[38] Such evidence points to a significant
advance at a certain juncture in history when the modern human mind –
with capabilities far beyond those of our nearest animal relatives – was
already clearly, intensely, and impressively at work. Philosopher of science
Holmes Rolston III has called this point in evolutionary history the "mind's
big bang."[39] From this point, human ideational, linguistic, and symbolic
capacities eventually headed us toward advanced culture-building that
included art, literature, mathematics, science, and many other human
achievements. Rolston also emphasizes two other key aspects of distinctively
human mental life: self-reflexiveness (second-order awareness of one's own
inner states) and the attribution of mind and inner states to others. Other
animals, he notes, have no "theory of mind."

Paleoanthropologists (who draw from paleontology, biological anthropology,
and cultural anthropology) are actually talking of "human uniqueness" – and
they marshal evidence to justify using this terminology. As anthropologist
Terrence Deacon states,

> Human consciousness is not merely an emergent phenomenon; it epitomizes
> the logic of emergence in its very form. Human minds, deeply entangled in
> symbolic culture, have an effective causal locus that extends across continents
> and millennia, growing out of the experiences of countless individuals.
> Consciousness emerges as an incessant creation of something from nothing, a
> process continually transcending itself. To be human is to know what it feels
> like to be evolution happening.[40]

The famous biologist Theodosius Dobzhansky, a deeply religious person,
puts it this way: "The biological evolution has transcended itself in the
human 'revolution.'"[41]

---

[38] Perhaps the most exciting work being done today on cultural evolution is at the
Çatalhöyük site in Turkey, which dates from the Neolithic period. See the Çatalhöyük
Research Project, www.catalhoyuk.com.

[39] H. Rolston III, *Three Big Bangs: Matter-Energy, Life, Mind* (New York: Columbia University
Press, 2010).

[40] T. Deacon, "The Hierarchic Logic of Emergence: Untangling the Interdependence of
Evolution and Self-Organization," in B. Weber and D. Depew (eds.), *Evolution and Learning:
The Baldwin Effect Reconsidered* (Cambridge MA: MIT Press, 2003), 306.

[41] T. Dobzhansky, *The Biology of Ultimate Concern* (New York: American Library, 1967), 58.

Scientists working in genetics these days have also detected what they claim are marks of human uniqueness. Craig Venter, whose Celera Genomics Project contributed significantly to the complete sequencing of human DNA by augmenting the federal government's Human Genome Project, emphasizes that the findings point to uniqueness. In reporting on the completion of Celera's genome work, Venter and his coauthors caution that materialist and reductionist conclusions cannot be drawn from what we now know of the human genome:

> In organisms with complex nervous systems, neither gene number, neuron number, nor number of cell types correlates in any meaningful manner with even simplistic measures of structural or behavioral complexity . . . . Between humans and chimpanzees, the gene number, gene structures and functions, chromosomal and genomic organization, and cell types and neuroanatomies are almost indistinguishable, yet the development modifications that predisposed human lineages to cortical expansion and development of the larynx, giving rise to language, culminated in a massive singularity that by even the simplest of criteria made humans more complex in a behavioral sense . . . . The real challenge of human biology, beyond the task of finding out how genes orchestrate the construction and maintenance of the miraculous mechanism of our bodies, will lie ahead as we seek to explain how our minds have come to organize thoughts sufficiently well to investigate our own existence.[42]

Ironically, Venter knows intimately the major scientific achievement that conclusively established the continuity between all living species at the genetic level and yet construes the findings in a larger context as evidence of human discontinuity in those qualities and behaviors that are distinctively human.

It is reasonable to say that this brief review of the state of the sciences provides a mixed report on human uniqueness. Continuities, obviously yes – but major discontinuities, yes indeed. Naturalist philosophers interpret the scientific data as resulting from small accidents in natural history such that humans cannot be "superior creatures" by some objective standard – as Dawkins, Simpson, and others say. Theistic philosophers interpret the data to suggest that finite embodied rational consciousness is a gift reflecting an infinite rational being, such that what we know biologically as *Homo sapiens*

---

[42] J. C. Venter et al., "The Sequence of the Human Genome," *Science*, 291 (2001), 1304–1351.

might also be known theologically as *Homo divinus* – a creature related to the divine. In fact, theologians as well as many scientists agree that consciousness of the divine must be on the list of essential human traits. Wentzel van Huyssteen, an expert in theology and science, believes that the emergence of human "self-consciousness" then precipitated consciousness of the divine. One piece of evidence for this would be in the spectacular cave paintings in France and Spain from the Upper Paleolithic.[43] Of course, naturalist philosophers and scientists who grant that religion aided survival in prescientific cultures now generally categorize religion as obsolete, anti-intellectual, and perhaps dangerous, matters we discuss further in Chapter 9.

We note that the preceding discussion largely treats the concept of the divine image in humans as something ontological – a specific reality with special intrinsic properties – but that relational and eschatological interpretations of human nature are also available. These interpretations typically follow twentieth-century neoorthodox theologian Karl Barth in saying that it is not the inherent capacity for relationship with God but the relationship itself that is the *imago Dei*. For Barth, the image of God is not constituted by anything humans *have* or *are* or *do* but rather by the gift of God's love.[44] Theologian Wolfhart Pannenberg, who takes science and anthropology very seriously, connects human nature and the *imago Dei* through the idea of "exocentricity" – that inner disposition toward self-transcendence that reaches beyond everything we experience in this world. This reach naturally interacts with biological instincts to express a longing for complete fulfillment – as seen in the evidence compiled by paleoanthropologists for the human "quest for meaning" or "search for religious fulfillment." According to Pannenberg, this longing leads us dynamically toward the eschatological future in which God brings finite humans into their destiny with him.[45] Regardless of whether ontological or relational interpretations are in focus, it is clear that, as van Huyssteen has remarked, all relevant theological reflection on the nature of humanity going forward must necessarily be done in interaction with multiple scientific disciplines.[46]

---

[43] J. W. van Huyssteen, *Alone in the World?: Human Uniqueness in Science and Theology* (Grand Rapids: Eerdmans, 2006), 164 ff.

[44] K. Barth, *The Doctrine of Creation* (Edinburgh: T. & T. Clark, 1958), 186 ff.

[45] W. Pannenberg, *What Is Man? Contemporary Anthropology in Theological Perspective* (Philadelphia, PN: Fortress Press, 1970), 7 ff.

[46] Van Huyssteen, *Alone in the World?*, 34 ff.

Evolution and Human Cognitive Powers

Darwin believed that "*[r]eason* stands at the summit" of all human powers, although he saw it as differing only in degree from the intelligence of animals.[47] Disagreeing with this perspective, many thinkers argue that the fact that rational human beings form *beliefs* and seek *knowledge* makes us noetic beings of a special sort. Interpretations of what knowledge is vary greatly – from Plato's objectivist view of knowledge as "justified true belief" to more subjectivist and pragmatic views of knowledge offered by contemporary naturalist philosophers who ally themselves with science. Churchland contends that the adaptation of our behavior as well as the neurophysiology producing our behavior are primarily to get the human physical organism moving appropriately (fleeing, fighting, etc.) and thus to have greater chances of survival. For her, beliefs are simply part of this evolutionary function.

Since objectively false beliefs in this light may be just as adaptive as objectively true beliefs, the question arises regarding whether evolutionary science undermines the claim that human beings have the special capacity to form objectively true beliefs and to acquire knowledge.[48] Although Churchland thinks that truth-as-adaptive-belief can replace more realist and objectivist theories of belief, analytic philosopher Alvin Plantinga maintains that naturalistic evolutionary accounts of the human knowledge-seeking process are incoherent. In debate with Daniel Dennett at the American Philosophical Association meeting in Chicago in 2009, Plantinga argued that the antecedent probability that cognitive powers capable of delivering objective truth (more truths than falsehoods) would arise in a universe described by naturalism is quite low.

Plantinga developed his much-discussed "evolutionary argument against naturalism" (EAAN), which contends that one cannot rationally accept both evolutionary theory and philosophical naturalism. As Plantinga articulates the argument, where R is the proposition that our cognitive faculties are reliable, N is the proposition that naturalism is true, E is the proposition that humans evolved through the natural selection of adaptive behavior, and P is probability, we get the following:

[47] Darwin, *Descent of Man*, 75.
[48] P. Churchland, "Epistemology in the Age of Neuroscience," Journal of Philosophy, 84, no. 10 (1987), 548.

1. $P(R/N \& E)$ is low.
2. One who accepts $N \& E$ and also sees that 1 is true has a defeater for R.
3. This defeater can't be defeated.
4. One who has a defeater for R has a defeater for any belief he or she takes to be produced by his or her cognitive faculties, including $N \& E$.

Therefore,

5. $N \& E$ is self-defeating and can't rationally be accepted.

Plantinga claims that this argument supplies a "defeater" for belief in evolutionary naturalism. In contemporary epistemology, a defeater is a proposition that has enough traction with a person – enough probative force – that it prevents rational belief in another proposition. Here, the putative defeater is implied by the conjunction of naturalism and evolutionary theory ($N \& E$): there are no grounds to think that rationality is reliable to deliver any truth whatsoever – including the truth of naturalist philosophy, adaptation theory, or whatever. Thus, according to Plantinga, the conjunction $N \& E$ must be rejected. Further, since evolutionary theory (E) is well confirmed, we have a strong reason to reject naturalism (N) because it cannot adequately explain why human rationality is reliable for truth and not just for survival.

In his public response, Daniel Dennett defended his evolutionary naturalism as follows:

> It is precisely the truth-tracking competence of belief-fixing mechanisms that explains their "adaptivity" in the same way that it is the blood-pumping competence of hearts that explains theirs. Hearts are for circulating the blood and brains are for tracking the relevant conditions of the environment and *getting it right*.[49]

It is difficult to understand how Dennett is giving a properly insightful reply, since for him to give an account of how natural selection was involved in developing the human capacity to recognize truth already rests on the assumption that his rationality is adequate to generate such an account. Dennett's account is hardly the requested grounds for why rationality is trustworthy for truth in the first place.

---

[49] D. C. Dennett and A. Plantinga, *Science and Religion: Are They Compatible?* (New York: Oxford University Press, 2011), 52.

Many theists maintain that EAAN leads to an argument *for* theism because, in a theistic universe, a rational God can will that rational creatures come forth, even if by a long evolutionary process. However, given that metaphysical naturalism asserts that the whole universe came ultimately from chance and runs by purely material processes that have no rationality or intentionality, many find it difficult to find in naturalism anything but a tiny likelihood that evolution would produce rational powers reliable for truth. Philosopher, linguist, and cognitive scientist Noam Chomsky agrees, stating that any congruence between the world and human belief-forming capacities "is just blind luck."[50] In this controversy, it is obvious that the bearing of the biosciences on our understanding of human nature will continue to be a point of much discussion and debate. Furthermore, the explanation of rational thought and all it entails will continue to be a major factor in deciding whether humans are unique and, if so, in what precise sense.

[50] N. Chomsky, *Language and the Problems of Knowledge* (Cambridge, MA: MIT Press, 2001), 157.

7   Love and Altruism in Biology and Religion

Cooperation and altruism are scattered broadly throughout nature, as sciences such as field biology and ecology have abundantly documented. Furthermore, most major religious traditions teach that regard for the other is reflective of the divine in nature broadly and should be central to human life. In pondering these two facts, a number of important issues arise. On the science side, although Darwinian principles appear to account for the abundant empirical evidence of mutualistic, symbiotic, integrative relationships in the living world, altruism is particularly difficult to explain in terms of the core principle of natural selection, which implies that animals retain only those traits that confer their own reproductive success or that of their close relatives. On the religious side, altruism and love are extolled as central to what all of life is about – and altruism and cooperation in the animal kingdom are interpreted as symbolic of this theme in creation.

In this chapter, we explore the problem of altruism in biology and related fields in interaction with religious teachings and insights on the topic. Among the fascinating questions we treat are the following: Has science shown that "selfish" needs preempt the possibility of genuine altruism, even in humans, or are there scientific reasons for believing that genuine altruism can and does exist? Can genes that blindly replicate themselves give rise to a level of function that transcends to some degree the inherent drive for replication? How does the scientific study of altruism in nonhuman species relate to altruism in humans, if at all? How does religion as a feature of culture influence altruistic behavior in humans? Do the perspectives of biology and religion on the nature of altruism ultimately diverge, as they initially seem to do, and thus produce irreconcilable frameworks for explaining altruism? Or might biology and religion be reconciled, say, by allowing findings in biology to enrich religious understandings, and by allowing religious insights to enlighten the facts of altruism in nature, including among humans? Furthermore, how do the fields of neuroscience and even primatology contribute to the interdisciplinary discussion of altruism?

The Central Theoretical Problem

For many decades, altruism has been described as the "central theoretical problem" in the application of biology to social behavior, resulting in a number of explanations for why altruism occurs in the living world. After all, the twin pillars of Darwinian natural selection are "struggle for survival" and "survival of the fittest," and in Darwin's Victorian England, the problem of squaring love with Darwinian principles was not lost on sensitive, thoughtful people. In 1850, Alfred Lord Tennyson grappled with the apparent conflict between Charles Lyell's geological findings of systemic violence and predation in wild nature and the Christian affirmation of love as the meaning of creation. Tennyson's poem *In Memoriam* contains these lines in canto 56:

> Who trusted God was love indeed
> And love Creation's final law
> Tho' Nature, red in tooth and claw
> With ravine, shriek'd against his creed

Earlier lines of the poem refer to paleontological digs that reveal that animal predation and death have occurred throughout evolutionary history and have led to the extinction of whole species. Is love really creation's final law, as Christianity teaches? Tennyson took nature as shrieking against this love. Thus, he asks in the poem whether humans – who have trusted in divine love and struggled for truth and justice – might ultimately experience the same fate as all extinct species. Is, then, humanity meaningless? Or is there meaning to be revealed in an afterlife, "behind the veil"?

Tennyson engaged the question of how to reconcile the shocking geological evidence – which Darwin later explained in 1859 – with traditional religious teachings on God and love. But Tennyson was not aware of the abundant evidence of altruism in the living world that has come to light in the twentieth century. The evidence of biological altruism creates a challenge for evolutionary science, but it also creates a challenge for religion to articulate the theological meaning of altruism and love in a created world run by evolutionary processes. For both science and religion, then, the key question raised is not why evil occurs, the topic of Chapter 4, but why good occurs – in this case, altruistic, self-giving behavior.

Some preliminary definitions and distinctions will allow us to join the discussion as it occurs in evolutionary biology. Altruism has a technical definition in evolutionary biology: an organism is said to behave altruistically when its behavior benefits other organisms at a cost to itself. We measure the costs and benefits in terms of reproductive fitness, which is the number of offspring. Altruistic behavior, then, reduces the number of offspring of the organism while increasing the offspring of others. In this context, it is only behavior that is classified as altruistic, not inner states like intentions. Indeed, altruistic behavior has been found in creatures that are not capable of conscious thought, such as insects. Thus, biological discussions of altruism focus on the consequences of the behavior rather than on the intentions driving the behavior. In social insects such as ants and bees, sterile workers care for the queen, feeding her and tending to her larvae, greatly enhancing her reproduction at their own expense. The list of altruistic behaviors in many species, particularly in species with complex social structures, is very long.

The puzzle immediately presents itself: natural selection suggests that organisms will behave in ways that increase their *own* chances of survival and reproduction, not the chances of others. When an animal behaves altruistically, it reduces its own fitness and is thus at a selective disadvantage compared to an animal that behaves selfishly. To envision the logic, imagine a Belding's ground squirrel, a small brown, furry creature with a short tail, which calls an alarm to warn the group of a predator at the risk of giving away its own location and being attacked. If the squirrel selfishly refuses to give an alarm call, it reduces its own chance of being attacked, and thus ostensibly has a selective advantage, while benefiting from the calls of others. We would normally conclude that natural selection would favor those ground squirrels that do not give alarm calls over those that do. So, why has alarm-calling behavior evolved in the first place and not been eliminated long ago by natural selection? On the surface, Darwinian principles seem at odds with this phenomenon.

Yet even from the beginning, Darwin sensed that this dichotomy was a false one. When organisms are social and closely related to one another, helping relatives reproduce has a probability of favoring the transmission of one's own genetic variation, albeit indirectly:

The subject well deserves to be discussed at great length, but I will here take only a single case, that of working or sterile ants . . .. [I]f such insects had been social, and it had been profitable to the community that a number should have been annually born capable of work, but incapable of procreation, I can see no very great difficulty in this being effected by natural selection.[1]

Here we see Darwin's thinking: that genes (or as he understood it, "heritable variation") could be transmitted by a social, related community and that natural selection could favor mechanisms for that means of transmission. Even though the cost to an individual's fitness might be absolute (i.e., sterility), the fitness of the group could be increased. Though later work by others would expand this nascent insight, Darwin had a rudimentary understanding of kin selection – natural selection acting on a group of relatives rather than on individuals alone.

Kin Selection and Reciprocal Altruism

Various evolutionary explanations have been proposed to account for altruism in the natural world. Although cooperation is widely observed in organisms of all types, from insects to mammals and even cells, it is frequently construed as "selfish" because there are mutual benefits to all individuals or genes involved.[2] Yet many other cases of cooperation resist simple explanation based on selfishness and have been the subject of several ingenious explanations that still appeal to fundamental Darwinian principles. In this neo-Darwinian engagement with these issues, the most important theories are *kin selection*, *reciprocal altruism*, *indirect reciprocity*, and *costly signaling*. In the present section, we particularly focus on kin selection and reciprocal altruism, which are often considered the two primary mechanisms for the evolution of social behaviors.

Kin selection, as we have seen, is natural selection acting collectively on a group of related organisms. The key concept for kin selection is that of "inclusive" fitness rather than individual fitness. Inclusive fitness, as the name suggests, goes beyond individual (so-called direct) fitness to include the effects that an individual has on the reproduction of its relatives ("indirect" fitness). Thus natural selection can favor situations where direct fitness

---

[1] C. Darwin, *On the Origin of Species* (London: Murray, 1859), 236.
[2] R. Dawkins, *The Blind Watchmaker* (New York: Norton, 1986), 380.

is low as long as indirect fitness compensates. Indirect fitness, of course, depends on how closely related two organisms are. On average, full siblings share one-half of their alleles, whereas cousins only share one-eighth. As such, there is a higher indirect fitness realized when an individual aids the reproduction of a closer relative. As biologist J. B. S. Haldane quipped in 1955 when asked whether he would risk his life to save a drowning person, "No, but I would do it for two brothers or eight cousins."[3]

In 1964, evolutionary biologist William Hamilton set forth the argument mathematically, demonstrating that an allele contributing to an altruistic behavior will be favored by natural selection under the following condition, known as *Hamilton's rule*: $b > c/r$. Here c is the cost incurred by the altruist organism; b is the benefit received by the recipients of the altruistic behavior; and r is the *coefficient of the relationship* between donor and recipient. Thus, the measure of the behavior is rendered in terms of inclusive fitness. Of course, the coefficient of relationship is the probability that the altruist and recipient share alleles at a given locus that are "identical by descent" – which means that they are copies of a single gene inherited from a shared ancestor. The greater the value of r, the greater the probability that the recipient of the altruistic behavior will also possess the allele contributing to the altruistic behavior. For full siblings, $r = \frac{1}{2}$; for cousins, $r = \frac{1}{8}$; and so on. Hence the humor in Haldane's quip. Hamilton thus concluded that genetic variation influencing altruism can indeed spread by natural selection if the cost incurred by the altruistic organism is offset by a compensatory benefit to sufficiently close relatives. Darwin's insight now had mathematical rigor.[4]

Though ample empirical evidence has long confirmed the prediction of Hamilton's rule that animals are more likely to display altruistic behavior toward their close relatives than toward more distantly related members of their species, detailed tests of Hamilton's rule are experimentally difficult. A major confounding factor is that altruistic behavior is commonly found in social organisms. Sociality makes teasing out the precise values of b and c challenging, since many factors in a social species influence these values. One notable study in red squirrels – an asocial mammal that, nonetheless,

---

[3] See Haldane's legendary quip on kin selection as reported in M. Nowak, *SuperCooperators: Altruism, Evolution, and Why We Need Each Other to Succeed* (New York: Free Press, 2011), 97.

[4] S. A. Frank, *Foundations of Social Evolution* (Princeton, NJ: Princeton University Press, 1998), 4, 46–47; D. C. Queller, "Quantitative Genetics, Inclusive Fitness, and Group Selection," *American Naturalist*, 139 (1992), 540–558.

exhibits altruistic behavior – provided an opportunity to test Hamilton's rule directly. Female red squirrels are known to adopt orphaned pups, though adding pups to a litter has a direct fitness cost to their own offspring. The average cost value (c) for adding an additional pup to a litter is known from long-term observations in this population. Similarly, r values are known either through genealogical observations or through DNA sequencing in some cases. Over a 19-year span, five adoptions were observed in the study area. In each case, Hamilton's rule was satisfied: females were willing to adopt only pups that increased their inclusive fitness. As such, this study provides rare experimental evidence that altruistic behavior – in this case, adopting an orphaned relative – can indeed be favored by natural selection as Hamilton proposed.[5] Of course, explaining altruism in terms of selfish genetic interests that produce a net gain to inclusive fitness still leaves altruism toward nonkin unexplained – creating a conceptual challenge to which we now turn.

In the early 1970s, evolutionary biologist Robert Trivers first proposed *reciprocal altruism* as the other major mechanism – alongside kin selection – that could promote the evolution of social behaviors in species.[6] The two theories share the same assumption: that some behaviors can be dictated by genes and can therefore be transmitted to future generations via natural selection. The gene-centered basis of nonkin reciprocal altruism is straightforward: it may pay one organism to help another organism. The theory maintains that an organism can act in such a manner that temporarily reduces its personal fitness while increasing the personal fitness of another organism, with the expectation that the other organism will act in a similar manner in the future. In common parlance, this is a scenario in which "if you scratch my back, I'll scratch yours." Thus, the initial cost of the helping behavior is offset by the likelihood of the returned helping behavior later – a transaction that would be sufficient for altruism toward nonkin to evolve. For the field of sociobiology, which seeks a "gene's-eye-view" for understanding social behavior, reciprocal altruism, along with inclusive fitness, is taken as foundational.

---

[5] J. Gorrell et al., "Adopting Kin Enhances Inclusive Fitness in Asocial Red Squirrels," *Nature Communications*, 1 (2010), art. 22.

[6] R. L. Trivers, "The Evolution of Reciprocal Altruism," *The Quarterly Review of Biology*, 46, no. 1 (1971), 35–57.

Prisoner's Dilemma

		Player 2	
		Altruist	Selfish
Player 1	Altruist	10,10	0,20
	Selfish	20,0	5,5

Figure 7.1 Prisoner's Dilemma

Two considerations complete our survey of reciprocal altruism. First, reciprocal altruism requires that individuals must interact more than once, have the ability to recognize other individuals with whom they have interacted, and discriminate between "cooperators" and "cheaters." Humans, of course, have these abilities in spades, but evidence for such abilities in other primates, bats, birds, and cetaceans (whales, dolphins, and porpoises) has been put forward as well, though not without controversy.[7] Cheaters eventually undermine their self-interest and will be punished by others refusing to help them or by the actions of a third party. Second, reciprocal altruism has been studied in game theory to provide the mathematics of how cooperation could have evolved when natural selection predicts selfish behavior. The famous Prisoner's Dilemma elucidates the benefit of cooperation between unrelated individuals.[8] The game consists of two players that may act altruistically (A) or selfishly (S) – and the options for their interaction form a matrix, with four possible outcome states for each player. If either player "defects" and chooses to act selfishly (S) when the other player chooses to act altruistically (A), the defector receives a large benefit at the expense of the altruist. If both choose to act selfishly (S), their benefit is drastically reduced. However, if both choose to act altruistically (A), they both receive a sizable benefit. Consider the example matrix in Figure 7.1.

A "one-shot" game of Prisoner's Dilemma naturally predicts that (S) is the winning strategy, as it produces either a modest reward if the other player also chooses (S) or a large reward if the other player chooses (A). Choosing (A)

---

[7] T. Clutton-Brock, "Cooperation between Non-Kin in Animal Societies," Nature, 462 (2009), 51–57.

[8] The Prisoner's Dilemma was introduced by American mathematician Albert W. Tucker to help illustrate a payoff matrix devised by Melvin Dresher and Merrill Flood. A. W. Tucker and P. D. Stranffin Jr., "The Mathematics of Tucker: A Sampler," The Two-Year College Mathematics Journal, 14, no. 3 (1983), 228–232. See a broad discussion of the game's significance in Robert Axelrod, The Evolution of Cooperation (New York: Basic Books, 1984).

in a one-shot game is risky, as it may result in no reward at all. Yet if the game is played repeatedly by the same pair – the "iterated" Prisoner's Dilemma – other strategies emerge. In one such strategy, "Tit-for-Tat," each player adjusts his or her behavior depending on the opponent's behavior in previous rounds. It has been demonstrated that the strategy of cooperating rather than defecting is significantly more successful when there is high probability of subsequent encounters, which are essentially statistical correlations between altruists and recipients.[9]

Two simple background assumptions of the model for biological studies are that reproduction is asexual and that type is perfectly inherited, such that altruistic organisms make altruistic organisms and selfish organisms make selfish organisms. Clearly, then, the altruistic type will be favored by selection only if there is a statistical correlation between partners. Otherwise, if the pairing is purely random with the same probability of having a selfish partner, evolutionary dynamics will favor selfishness, which will increase in frequency in each successive generation.[10] For the development of social behavior in a population over the long term, as evolutionary mathematician Martin Nowak has confirmed, repeated interactions of cooperators or altruists that can recognize one another are required, which is not the scenario of the one-shot Prisoner's Dilemma.[11] These implications are borne out in the iterated Prisoner's Dilemma, where players adjust their behavior dependent upon what their opponent has done in previous rounds.[12] In frequent encounters, for example, "cheaters" that have received help but then refused to give it are identified and receive no further help. Additionally, third-party punishment for cheating may come into play.

Although reciprocal altruism has much theoretical validation, the number of empirically documented instances among nonhuman animals is relatively small. Biologist Gerald Wilkinson's studies of the evolution of social behavior have revealed that blood sharing in vampire bats may provide

[9] S. Maynard, "The Origin of Altruism," *Nature*, 393 (1998), 639–640. See also S. Okasha, "Genetic Relatedness and the Evolution of Altruism," *Philosophy of Science*, 69, no. 1 (2002), 138–149.

[10] B. Skyrms, *Evolution of the Social Contract*, 2nd edn. (Cambridge, UK: Cambridge University Press, 2014).

[11] M. Nowak, *Evolutionary Dynamics: Exploring the Equations of Life* (Cambridge, MA: Harvard University Press, 2006).

[12] R. Axelrod and W. D. Hamilton, "The Evolution of Cooperation," *Science*, 211, no. 4489 (1981), 1390–1396.

one empirical example. Bats donate blood by regurgitating to other members of their group that failed to feed on any given night, since these bats die if they are food-deprived for only a few days.[13] Of course, one possibility is that this is merely an example of kin selection if recipients tend to be close relatives of the donor. Recent work, however, has shown that reciprocity is a better predictor of food sharing than relatedness in this species,[14] and as such it remains one of the best-supported examples of nonhuman reciprocal altruism even as other proposed cases have increasingly been suggested to have alternative nonaltruistic explanations.[15]

Although our understanding of altruistic social behavior in terms of Darwinian principles has advanced a long way, interesting conceptual issues and questions remain. Some would say that the explanations we have surveyed – kin selection and reciprocal altruism – are based on indirect ways of promoting selfish interests. As the critique goes, the behaviors in question turn out to be only "apparently" altruistic and not "really" altruistic because they are actually performed for the benefit of the actor as strategies of pursuing its own inclusive fitness benefit. The concept of the "selfish gene" made famous by Richard Dawkins states that these strategies are actually "selfish" because the behavior of genes is to replicate and increase their representation in the gene pool.[16] Obviously, this scientific answer does not employ the term "altruism" in evolutionary biology as synonymous with the same term in vernacular usage in which there are conscious, other-regarding intentions, particularly since the majority of living creatures do not seem capable of inner intentions in the same way humans conceive of them. Thus, theories of biological altruism use the term "altruism" defined in terms of inclusive fitness consequences rather than conscious intentions.

Perhaps the most interesting question concerns whether theories of biological altruism apply to humans. In the 1980s, sociobiology created an ongoing controversy by its assertion that ideas pertaining to the evolution of animal behavior can be extrapolated to *Homo sapiens* as an

[13] G. S. Wilkinson, "Reciprocal Food Sharing in the Vampire Bat," *Nature*, 308 (1984), 181–184.
[14] G. G. Carter and G. S. Wilkinson, "Food Sharing in Vampire Bats: Reciprocal Help Predicts Donations More Than Relatedness or Harassment," *Proceedings of the Royal Society, Series B*, 280, no. 1753 (2013), https://royalsocietypublishing.org/doi/10.1098/rspb.2012.2573.
[15] Clutton-Brock, "Cooperation between Non-Kin in Animal Societies," 51–57.
[16] R. Dawkins, *The Selfish Gene* (New York: Oxford University Press, 1976).

evolved species.[17] Clearly, some altruistic human behavior that occurs is predicted by kin selection (e.g., raising orphaned children who are close relatives) and by reciprocal altruism (e.g., helping nonkin who have previously helped us). It is also the case that humans tend to exhibit altruistic behavior preferentially toward close relatives. On the other hand, humans cooperate far more with nonkin than do other species and seem to act from conscious motivation for the welfare of others – thus seeming to exhibit more "real altruism" than other species do. These things are true even when biological fitness is not affected in terms of cost and benefit. Ernst Fehr and Urs Fischbacher have shown that, while cooperative behavior increases when kin selection, reciprocal altruism, indirect reciprocity, and costly signaling are at stake, there is also widespread evidence that humans clearly cooperate when such evolutionary behaviors are apparently not at stake.[18] Some efforts – by William Hamilton and others – seek to explain distinctively human altruism by employing gene-centered models, particularly kin selection because of its successful theoretical extensions and substantial empirical support. One challenge that arises, of course, is extending altruistic behaviors to those who are not close kin. We will discuss these topics further as we proceed, but it is clear that these are all still causal theories of morality, which are not new in the discussion of morality.

The Group beyond (Close) Kin

Although kin-selection theory has long been the primary evolutionary framework for explaining prosocial behavior, with its salient prediction that positive social behavior should have a high level of correlation with relatedness, group selection (also known as multilevel selection) has also received considerable attention. Multilevel selection is, as it sounds, the observation that selection may act at higher orders of organization than merely the

[17] R. Boyd and P. J. Richerson, "Culture and the Evolution of the Human Social Instincts," in N. J. Enfield and S. C. Levinson (eds.), *Roots of Human Sociality: Culture, Cognition, and Interaction* (New York: Berg, 2006), 453–476; see also S. Bowles and H. Gintis, *A Cooperative Species: Human Reciprocity and Its Evolution* (Princeton, NJ: Princeton University Press, 2011).

[18] E. Fehr and U. Fischbacher, "The Nature of Human Altruism," *Nature*, 425 (2003), 785–791.

individual or close relatives.[19] If selection acts on a group relative to another group (e.g., on variation that is relevant to intergroup conflict), then direct fitness costs to individuals may be compensated for by average indirect fitness gains to the group as a whole. As we have discussed, indirect fitness is the key driver for kin selection – meaning that multilevel selection can be seen as equivalent to kin selection and describing the same biological process but at a wider scale. In nature, groups tend to be more closely related, on average, to in-group members than to those outside the group. Though there is some debate among biologists on this point, the consensus at present seems to favor their equivalence.[20] Although we cannot pursue the debate here, we observe that there is at least general agreement that group selection does illuminate the differential survival and reproduction of groups. As such, the theory is a fascinating candidate for a potential solution to the central problem of sociobiology because it asserts that even very costly social behaviors can be favored by natural selection – as long as the direct costs are outweighed by a sufficient amount of indirect benefit to individuals that are not close kin in the kin-selection sense but are, nonetheless, related and part of the group. The group would thus have a greater-than-average probability of sharing the actor's alleles, including the alleles that promote the social behavior in question.[21]

In primates, prosocial behavior may have its origins in cooperative breeding, a reproductive strategy where infant-rearing tasks are shared among group members. In particular, allomaternal care – females caring for infants who are not their own offspring – correlates strongly with prosocial behavior for a large number of primate species. This suggests that human prosocial behavior, though it exceeds that of other species, is, nonetheless, a general feature of cooperatively breeding primates.[22] Despite this understanding, humans remain something of an "evolutionary puzzle"[23] in that they exhibit

[19] See S. A. Frank, "Natural Selection. VII. History and Interpretation of Kin Selection Theory," *Journal of Evolutionary Biology*, 26, no. 6 (2013), 1151–1184.

[20] J. Kramer and J. Meunier, "Kin and Multilevel Selection in Social Evolution: A Never-Ending Controversy? [Version 1; Peer Review: 4 Approved]," *F1000Research*, 5, no. 776 (2016), https://f1000research.com/articles/5-776.

[21] K. R. Foster, "A Defense of Sociobiology," *Cold Spring Harbor Symposia on Quantitative Biology*, 74 (2009), 403–418.

[22] J. Burkart, O. Allon, F. Amici, et al., "The Evolutionary Origin of Human Hyper-Cooperation," *Nature Communications*, 5, no. 1 (2014), art. 4747.

[23] E. Fehr and S. Gächter, "Altruistic Punishment in Humans," *Nature*, 415 (2002), 137–140.

high levels of cooperation beyond this baseline, even with genetically unrelated strangers whom they will never encounter again:

> Human altruism goes far beyond that which has been observed in the animal world. Among animals, fitness-reducing acts that confer fitness benefits on other individuals are largely restricted to kin groups .... If we randomly pick two human strangers from a modern society and give them the chance to engage in repeated anonymous exchanges in a laboratory experiment, there is a high probability that reciprocally altruistic behavior will emerge spontaneously.[24]

Proposed explanations for the depth of human altruism are diverse. Reputation building, direct reciprocity, indirect reciprocity, costly signaling, and avoiding punishment by third parties all have experimental support in humans and appear to be part of the mechanisms that make humans exceptional in this regard.[25] Group selection in humans casts a far wider net than for any other species, not least because of our complex cognitive abilities and long evolutionary history of cooperation.

Although both individual and group selection theories explain very well the adaptive advantage of intragroup prosocial behavior among many animal species, including intragroup sacrifice that is unreciprocated, the challenge persists of explaining extreme forms of altruism found among humans, which would include sacrifice for nongroup members (even competitors and enemies) that brings no reproductive advantage. Adding to the challenge of explaining extreme altruism in humans are the testimonies of those who experience a sense of fulfillment resulting from engaging in such personal sacrifices. Indeed, some moral systems and many religions actually recommend sacrificial altruism as a desirable and admirable life strategy. As an ethical doctrine, altruism holds that we are morally required to act in ways that benefit others. As a religious principle or teaching, altruism is often valued as the way of a happy life and of living in accord with the divine. Almost all of the world's developed religions project attitudes and actions that fall outside the explanatory frameworks that satisfactorily explain altruism among the other social animals.

---

[24] Fehr and Fischbacher, "The Nature of Human Altruism," 785–791.

[25] For a detailed review, see R. Kurzban, M. N. Burton-Chellew, and S. A. West, "The Evolution of Altruism in Humans," *Annual Review Psychology*, 66, no. 1 (2015), 575–599.

Many religions contain the well-known rule of reciprocity, often called the Golden Rule. In the sixth century BC, Siddhartha Gautama, who became the Buddha after his enlightenment, taught a key principle: "Treat not others in ways that you yourself would find hurtful."[26] The Buddha even developed meditative techniques aimed at developing an attitude of loving kindness (Sanskrit, *maitrī*) toward others. In this vein, the Dalai Lama has commented that, in order not to harm others, one must cultivate kindness and compassion toward others.[27] Also, in the sixth century BC, Confucius said in the *Analects*, "One word which sums up the basis of all good conduct is loving-kindness."[28] The New Testament records that Jesus stated, "In everything, do to others as you would have them do to you; for this is the law and the prophets."[29] Here we see a shift in emphasis from "do not harm" to a more positive expression – "do unto others." In the giving of the Great Commandments, Jesus emphasized love behind the positive actions:

> One of the scribes came up and heard them disputing with one another, and seeing that he answered them well, asked him, "Which commandment is the most important of all?" Jesus answered, "The most important is, 'Hear, O Israel: The Lord our God, the Lord is one. And you shall love the Lord your God with all your heart and with all your soul and with all your mind and with all your strength.' The second is this: 'You shall love your neighbor as yourself.' There is no other commandment greater than these."[30]

The second Great Commandment emphasizes love for neighbor, love for the other, referring to an inner state from which loving actions flow.

In fact, Christians see all New Testament teachings on love, taken together, as explaining important nuances of unconditional self-giving love – or *agape* in the original Greek of the New Testament. Jesus said, "No greater love hath a man than that he lay down his life for a friend."[31] But perhaps Jesus's most shocking statement is that we should love our enemies:

> You have heard that it was said, "Love your neighbor and hate your enemy." But I tell you, love your enemies and pray for those who persecute

---

[26] The Buddha, *Udana-Varga*, 5.18.
[27] The Dalai Lama and H. C. Cutler, *The Art of Happiness* (New York: Riverhead Books, 1998), 64.
[28] Confucius, *Analects*, 15.23.
[29] Matthew 7:12. Jesus is quoting from the Old Testament Leviticus 19:18.
[30] Mark 12:28–31.   [31] John 15:13.

you, that you may be children of your Father in heaven. He causes his sun to rise on the evil and the good, and sends rain on the righteous and the unrighteous. If you love those who love you, what reward will you get? Are not even the tax collectors doing that? And if you greet only your own people, what are you doing more than others? Do not even pagans do that? Be perfect, therefore, as your heavenly Father is perfect.[32]

These teachings on love position costly altruism in humans far beyond any forms of direct or indirect reciprocity and create a difficult problem for evolutionary biologists to solve. Thus, the search for a sufficient evolutionary explanation of human altruism continues, not only in biology but in the behavioral sciences, theology, and philosophy as well.

As we continue to seek understanding of altruism, we must sort through various biological theories, whether adaptationist or nonadaptationist, but we may well suspect that a sufficient understanding will have to be holistic, accounting for multilevel phenomena that operate at a higher level of scale than gene or group selection and considering whether these higher-level phenomena are related to but possibly transcend requirements of biological fitness. The particularly difficult and intriguing cases that seem to go beyond our biology are the human altruistic acts, typically encouraged or justified by religion, that include little or no possibility of reciprocity – such as giving to the poor, committing to a life of sacrificial service, or even submitting to martyrdom.

Sociobiology and Evolutionary Psychology

The thought that humans might be able to transcend their biological interests meets perhaps its most difficult and most systematically developed challenge in the field of sociobiology. Although the term "sociobiology" was used as early as the 1940s, E. O. Wilson, biologist and noted expert on ant populations, published *Sociobiology: The New Synthesis* in 1975, defining it as the "systematic study of the biological basis of all social behavior."[33] In context, the "biological basis" meant the social and ecological causes driving behavior in animal populations. This attempt to explain the evolutionary

---

[32] Matthew 5:43–48.
[33] E. O. Wilson, *Sociobiology*, 25th anniversary edn. (Cambridge, MA: Harvard University Press, 2000), 4.

mechanics behind social behavior – such as altruism and aggression – involved the principle that an organism's evolutionary success is tied to getting its genes represented in the next generation. Biologists were generally accepting of the project, but reception among sociologists was mixed. Some sociologists criticized Wilson's last chapter for prematurely extrapolating from nonhuman species to human societies, while other sociologists explicitly admitted that they could learn from biology. Due to intense interest, the "sociobiological wars" reached even the front page of the *New York Times* in a review of Wilson's book.

The common assumption of the pioneers and practitioners of sociobiology was that kin selection and reciprocal altruism (both direct and indirect) are completely sufficient explanations of human cooperative behaviors. The influential evolutionary biologist Richard Alexander categorically stated,

> Defining morality as self-sacrificing, failing to understand all aspects of indirect reciprocity, and injecting the question of motivation ... have caused crucial points to be missed .... [N]epotism and reciprocity ... account for human societal structure. To the extent that morality is a within-group cooperativeness generated in the context of between-group competition [it is] inconsistent with the principles typically associated with modern definitions of moral behavior.[34]

Alluding to Herbert Spencer's two codes of morality – the "code of amity" and the "code of enmity" – Alexander indicates that both forms of conduct were exposed to natural selection and became opposing aspects of the human mental repertoire. In this way, friendliness, love, altruism, charity, self-sacrifice, and "all the Christian virtues," which are universally approved – as well as envy, deceit, aggression, hate, and other antisocial traits, which are universally condemned by civilized societies – all increased the chances of survival in evolving humanity. The underlying theme is that humans do not come on the scene as blank slates but rather come highly programmed out of deep evolutionary history. At the time Alexander wrote, there were gaps in our knowledge about the neural basis of such qualities, but these gaps have been gradually closing in the intervening years, as we shall see later in this chapter.

[34] R. D. Alexander, *The Biology of Moral Systems* (1987; repr. London: Routledge, 2017), 194–196.

Thus, Wilson and most sociobiologists take all forms of cooperative (or reciprocal) behavior to be rooted in underlying selfishness, which strategically utilizes cooperation as a means to selfish ends. By way of analogy, we might think of the free-market economic theory of Adam Smith, the eighteenth-century father of modern economics, which asserts that, if humans pursue their natural self-interest in a free, ordered, competitive marketplace, they will naturally produce, all other things being equal, the most prosperous society.[35] But the basic point of sociobiology is that to understand our behavior, we must understand the underlying evolutionary forces. For example, scientists have reported finding genes that influence some social behaviors – such as parenting patterns in mammals and even the social feeding patterns in worms – which opens up a new subfield that we might call the "molecular biology of social behavior."

Among other points of controversy, critics tend to see sociobiology as describing the evolutionary forces shaping our social behavior as deterministic – in this case, biological or genetic determinism – without adequately allowing for environmental influences on human development. One question regards how exegetically accurate the criticism is to Wilson's own writing; another question regards how the enduring question of determinism and free will plays out within this new sociobiological context. We cannot pursue the exegetical question, but we do recognize the importance of the question of determinism. Some defenders of sociobiology, such as psychologist David Barash, argued that sociobiology does not entail that genes completely control behavior but rather claims that genes together with experience and culture contribute to behavior.[36]

Evolutionary psychology took a different direction from sociobiology partly because of the negative connotations associated with the term "sociobiology" and partly because of a shift in emphasis from the study of adaptive behavior to the study of psychological adaptations, such as the functional organization of the brain and even the social and moral emotions. Darwin himself seemed to forecast such a study in the last chapter of the *Origin*:

> In the distant future I see open fields for far more important researches. Psychology will be based on a new foundation, that of the necessary

---

[35] See Smith's remarks on the "invisible hand." A. Smith, *Wealth of Nations* (1776), 4.2.
[36] D. P. Barash, "The New Synthesis," *The Wilson Quarterly*, 1, no. 4 (1977), 108–120.

acquirement of each mental power and capacity by gradation. Light will be thrown on the origin of man and his history.[37]

Emerging in the 1980s, evolutionary psychology was given prominence in the 1992 book *The Adapted Mind: Evolutionary Psychology and the Generation of Culture*, edited by Jerome H. Barkow, Leda Cosmides, and John Tooby. The new field took a multidisciplinary (and potentially unifying) approach, weaving together evolutionary biology, human evolution, hunter–gatherer studies, cognitive science, neuroscience, and psychology to gain new insights into the human mind and the brain.[38]

According to evolutionary psychologists, the adaptive problems that our hunter–gatherer ancestors faced during the Pleistocene help explain the functional designs of the emotions, cognitive instincts, and motivations that human evolution produced. As humanity developed from small groups, which had to cooperate with others to obtain food, to larger, more complex societies, selection favored altruism and cooperation beyond one's relatives. Cooperative impulses were further favored by selection as humans encountered competition from other groups, which made them more "group-minded," identifying with others in their society with whom they were not directly acquainted. Cultural conventions and norms provided incentive structures for feelings of social interdependence and responsibility.

Furthermore, evolutionary psychologists say that the human brain has been shaped to be cooperative, compassionate, and community-minded for survival. They have postulated that the mind has specialized modules for such responses, arguing that domain-specific cognitive modules would have been selected because they enabled our ancestors to react quickly and effectively to environmental challenges in contrast to more general cognitive mechanisms that work more slowly and thus would have been eliminated in the course of evolutionary history. The idea that the mind may to some extent be composed of innate mental modules has been criticized by a number of cognitive scientists who refer to neurological studies showing "brain plasticity" – the ability of areas of the brain to take on different functions in response to experience and environment. We cannot pursue the extensive discussion on this issue between evolutionary psychologists

---

[37] Darwin, *Origin of Species*, 488.
[38] See J. H. Barkow, L. Cosmides, and J. Tooby (eds.), *The Adapted Mind: Evolutionary Psychology and the Generation of Culture* (New York: Oxford University Press, 1992).

and cognitive scientists, but we do note that researchers in the relevant sciences agree that humans have developed a "social brain" that has evolved for compassion, cooperation, and community, while they continue to study the details of its evolution and function.

The Neurobiology of Love

The scientific claim that altruism was built into humans during their evolutionary journey can be partially explained by reference to utilitarian consequences that enhanced group survival, but neuroscience supplements and perhaps transcends this explanation with information suggesting that much altruistic behavior is associated with inner states such as empathy, compassion, and love. Although nothing approaching the complexity of human love is observed in any other social animals, studies have isolated specific aspects of love defined operationally in animal populations for experiment and measurement, which may in turn provide direction for studying some sorts of love in humans. Once again we confront a familiar pattern: human biology is rooted in our evolutionary history but reaches an expression that is unique among animals in its depth and complexity.

Research into the neurochemistry of pair-bonding in mammals illustrates this general observation. Oxytocin and vasopressin are two neuropeptides that are key regulators of social and reproductive behaviors in a wide range of animals, including mammals. Both are ancient, predating the split between invertebrate and vertebrate lineages some 700 million years ago.[39] In socially monogamous mammals,[40] manipulating these neuropeptides can either favor or disrupt pair-bonding – a result most famously seen in studies of prairie voles in the 1990s. Continuing work on this species has provided evidence that epigenetic modification of the vasopressin receptor gene and oxytocin gene may contribute to pair-bond maintenance. After mating, chemical modification of the DNA of both genes causes an increase in their production in the nucleus accumbens – a region of the basal forebrain

[39] Z. R. Donaldson and L. J. Young, "Oxytocin, Vasopressin, and the Neurogenetics of Sociality," *Science*, 322, no. 5903 (2008), 900–904.
[40] In biology, a species can exhibit either *sexual* monogamy, where pair-bonded partners have sex exclusively with each other, or *social* monogamy, where partners pair to raise offspring but might also "cheat" with extra-pair sexual liaisons. Humans are a socially monogamous species, since not all human pair-bonds are sexually exclusive.

heavily involved in the dopaminergic reward pathway – in female voles, perhaps confirming and maintaining mate choice.[41] In humans, there is growing evidence that oxytocin levels influence romantic interactions[42] and that variation in the oxytocin receptor gene correlates with human prosocial behavioral differences.[43] However, as neuroscientist and psychiatrist Thomas R. Insel has cautioned, the human brain is more complex and "dominated by a massive cortex that governs the hypothalamus and other deep brain structures." Thus, the action of hormones "may modify human behavior, but due to the dominance of the cortex, intellectual, spiritual, and cultural influences ultimately may determine human attachments independent of hormonal state."[44]

Recent work with humans further underscores that human attachment and prosocial/altruistic behavior both have basal, subcortical components as well as higher-order, cortical involvement. Functional magnetic resonance imaging (fMRI) allows observation of brain activity in live subjects as they experience various phenomena, and such studies directed toward the question of human pair-bonding reveal both levels of activity. Notably, the subcortical dopaminergic reward pathway is activated in humans in pair-bonding situations, including the nucleus accumbens, as well as higher cortical regions involved with cognitive processing such as the prefrontal cortex.[45] A recent meta-analysis of fMRI studies comparing strategic giving, altruistic giving, and selfish actions found that both forms of giving activated the nucleus accumbens as well as several cortical regions associated with reward and value judgments, suggesting that they share neural circuitry. If so, altruistic, intrinsic motivation and reward in humans may be an

[41] H. Wang, F. Duclot, Y. Liu, et al., "Histone Deacetylase Inhibitors Facilitate Partner Preference Formation in Female Prairie Voles," *Nature Neuroscience*, 16, no. 7 (2013), 919–924.

[42] S. B. Algoe, L. E. Kurtz, and K. Grewen, "Oxytocin and Social Bonds: The Role of Oxytocin in Perceptions of Romantic Partners' Bonding Behavior," *Psychological Science*, 28, no. 12 (2017), 1763–1772.

[43] J. Li, Y. Zhao, R. Li, et al., "Association of Oxytocin Receptor Gene (OXTR) Rs53576 Polymorphism with Sociality: A Meta-analysis," *PLoS ONE*, 10, no. 6 (2015).

[44] T. R. Insel, "Implications for the Neurobiology of Love," in S. G. Post et al. (eds.), *Altruism and Altruistic Love: Science, Philosophy, and Religion in Dialogue* (New York: Oxford University Press, 2002), 262.

[45] S. Ortigue et al., "Neuroimaging of Love: fMRI Meta-analysis Evidence toward New Perspectives in Sexual Medicine," *The Journal of Sexual Medicine*, 7, no. 11 (2010), 3541–3552; Y. Cheng et al., "Love Hurts: An fMRI Study," *NeuroImage*, 51, no. 2 (2010), 923–929.

extension of extrinsic motivation for reward.[46] Neurobiology has revolutionized our understanding of how the physical brain processes such functions as memory, sensory input, and motor function and is currently making progress toward understanding the neural basis of altruistic love.

Although neuroscience continues to study the brain, it has already provided a more complete picture than we have ever had of how the brain processes the cognition required for social behavior. In biological terms, our brains have evolved to be innately and instinctively cooperative and compassionate. Actually, Darwin's view that empathy, not selfishness, lies at the basis of our social behaviors foreshadowed the converging contemporary scientific evidence for this theme. What Darwin called the "high moral rules" do not give rise to, but rather are founded on, what he called the "social instincts."[47] For him, these social instincts, such as "sympathy," are at the core of sociability and the forms of mutuality among the social animals, including risking life for others of the community.[48] A number of studies now document that prosocial, empathic acts benefit those performing the acts and not only the recipients. A study of volunteerism, for example, shows that volunteerism closely correlates with self-reports of greater happiness, health, and sense of well-being. In fact, compared with nonvolunteers, older adults who regularly volunteered had reduced rates of anxiety and depression and had greater physical health and longevity.[49] Similar studies abound, and it is entirely uncontroversial among scholars that altruistic behavior is beneficial to those who offer it as well as to those who receive it:

> When it comes to the pursuit of happiness, popular culture encourages a
> focus on oneself. By contrast, substantial evidence suggests that what
> consistently makes people happy is focusing prosocially on others.[50]

Interestingly, various moral and religious traditions consider compassion and kindness to be important to the religious life, not just as outward actions

[46] J. Cutler and D. Campbell-Meiklejohn, "A Comparative fMRI Meta-analysis of Altruistic and Strategic Decisions to Give," *NeuroImage*, 184 (2019), 227–241.

[47] C. Darwin, *The Descent of Man*, 2nd edn. (London: John Murray, 1874), 122.

[48] Darwin, *Descent of Man*, 2nd edn., 100–104.

[49] J. S. Barron et al., "Potential for Intensive Volunteering to Promote the Health of Older Adults in Fair Health," *Journal of Urban Health*, 86, no. 4 (2009), 641–653.

[50] S. K. Nelson et al., "Do unto Others or Treat Yourself? The Effects of Prosocial and Self-focused Behavior on Psychological Flourishing," *Emotion*, 16 no. 6 (2016), 850–861.

but as inward motivational states. Yet religious teachings differ over the exact nature of compassionate motivations. For instance, Buddhism speaks of the value of attitudes and actions that are compassionate but invests the concept of compassion with distinctively Buddhist meaning. The Buddha's enlightenment enabled him to see "things as they truly are" (yathā-bhū-tam), which is a total reality in which suffering is fundamental. Buddhism explains that Gautama was transformed into a completely "compassionate" person who taught that the "way of peace" is finding nirvana and then showing kindness and compassion to others. Compassion in the Buddhist sense is, then, not so much a feeling but rather a mental posture based in the transcendent knowledge that one is part of a larger, interdependent whole. His Holiness the Dalai Lama elaborates on the ancient doctrine:

> According to Buddhism, compassion is an aspiration, a state of mind, wanting others to be free from suffering. It's not passive – it's not empathy alone – but rather an empathetic altruism that actively strives to free others from suffering. Genuine compassion must have both wisdom and loving-kindness.[51]

Thus, from the very origin of Buddhism to the present, enlightenment and compassion are essentially linked.[52]

Christianity also teaches compassion based on its own theological vision. Jesus told the parable of the Good Samaritan who gave help to a man who had been beaten and robbed by thieves and left for dead alongside the road. A priest and a Levite had previously passed by but had consciously ignored the man, although each was under religious obligation, as stipulated in religious rules, to help people in distress. Later, a Samaritan came along and "had compassion" on the suffering man, who was a complete stranger, not related by blood or tribe. In this story, compassion was an inner prompt to action that was surely deeply rooted in social instincts. Since all persons have a wide variety of relationships – from intimate family to friends and strangers – Christianity has a long tradition, stemming from Augustine, on the "ordering of the loves" (ordo amoris).[53] Augustine even redefines the

---

[51] T. Gyatso (the 14th Dalai Lama), *Essence of the Heart Sutra: The Dalai Lama's Heart of Wisdom Teachings*, trans. G. T. Jinpa (Somerville, MA: Wisdom Publications, 2015), 49.

[52] R. L. F. Habito, "Compassion Out of Wisdom: Buddhist Perspectives from the Past toward the Human Future," in S. G. Post et al. (eds.), *Altruism and Altruistic Love*, 362.

[53] Augustine, *City of God*, 15.22.

classical virtues of justice, prudence, courage, and temperance as ultimately being ways of loving God and neighbor.[54]

At this point, we can confidently say that altruism plays a central role in biology in achieving reproductive success, but we also now know that altruism is associated with certain cognitive and emotional processes. The whole topic is clearly of interest to the behavioral sciences and to religion, theology, and philosophy as well. The fact that religion promotes (and therefore is a predictor of) prosocial behavior may be explained in different ways. From a religious perspective, religiosity is generally associated with an objective view of morality, which in turn emphasizes positive actions toward others. Yet religion also promotes inward motives such as love and compassion, typically emphasizing that the subjective states are more important than simply following rules. From a scientific perspective, neurobiology provides important information about the chemical and brain processes involved in the morally and religiously important subjective states of love and compassion. Whether the religious and neurobiological explanations are seen as dichotomous is very much dependent on worldview. At one extreme, thinkers holding a naturalist perspective take the science as adding one more piece to their causal and mechanistic explanation of the physical world. At the other extreme, the dichotomy is reinforced by religious believers that do not have helpful categories for engaging the science and may tend to overly spiritualize the development of prosocial feelings. Taking a more moderating position, the sophisticated theist can say that causal explanation and purposeful theological explanation are not dichotomous because God's material creation was intended to include entities and processes that mediate higher spiritual values. In Christianity, for example, the doctrines of creation, Incarnation, and sacraments imply a positive role for the physical.

Altruism and Empathy in Mammals

In regard to mammals, it is virtually uncontested at this point that various forms of altruism emerge from their biological architecture, which provides for feelings of sympathy with others in the group. Darwin put it succinctly:

---

[54] Augustine, *Catholic and Manichaean Ways of Life*, 1.25; *Letter* 155.12; cf. *Letter* 155.16.

The social instincts lead an animal to take pleasure in the society of its fellows, to feel a certain amount of sympathy with them, and to perform various services for them.[55]

Many philosophers who seek to naturalize ethics see in animal behavior the precursors of human social instincts and morality.

No one has made this point more effectively than primatologist Frans de Waal. In his book *Mama's Last Hug: Animal Emotions and What They Tell Us about Ourselves*, de Waal relates the story of the strong bond between Mama, an old dying chimp in captivity, and biologist researcher Jan van Hooff, whom Mama had known for 40 years. De Waal describes Jan entering the cage of the weak and dying chimp for their last visit together:

When Mama finally does wake up from her slumber ... she expresses immense joy at seeing Jan up close and in the flesh. Her face changes into an ecstatic grin, a much more expansive one than is typical of our species .... [S]he yelps – a soft, high-pitched sound for moments of high emotion .... [S]he ... reaches for Jan's head [and] gently strokes his hair, then drapes one of her long arms around his neck to pull him closer. During this embrace, her fingers rhythmically pat the back of his head and neck in a comforting gesture that chimpanzees also use to quiet a whimpering infant.[56]

Fortunately, a video was taken of this touching encounter for all to see.[57] In her long life, Mama was actually a matriarch, who had even been observed facilitating reconciliations between bellicose male chimps and even tenderly kissing a chimp that had been ostracized in order to coax him back to the group.

De Waal uses the story of Mama's last hug to communicate his larger message regarding how species that are related to humans exhibit emotions related to human emotions – anger, fear, affection, and the like – and show those emotions with similar body language. He notes that Darwin commented in *The Expression of the Emotions in Man and Animals* that other primates even employ humanlike facial expressions in emotionally charged situations, demonstrating impressive physical similarities with how humans display

---

[55] Darwin, *Descent of Man*, 2nd edn., 98.

[56] F. de Waal, *Mama's Last Hug: Animal Emotions and What They Tell Us about Ourselves* (New York: W. W. Norton, 2020), 13–14.

[57] Jan van Hooff posted video of the encounter to his YouTube channel, https://youtu.be/INa-oOAexno/.

emotions and thus strongly suggesting similarities in their inner lives.[58] De Waal has also argued that human morality can be naturalized by rooting it in the natural sympathy for others that is displayed among numerous animal species. He has even conducted and filmed various experiments showing the instinctive sense of fairness between monkeys, which seems to have an emotional basis – a finding we explore further in Chapter 8.

Scientifically, then, we can say that human altruistic love is grounded in empathetic affection – and that the building blocks of empathetic feelings are found in nonhuman species. We can further say that the human emotional capacity for empathy and compassion are elevated, refined, and directed by culture, particularly its moral and religious traditions. The moral teachings that a culture employs to enforce cooperation are widely judged to receive additional authority via religion, which contains powerful symbols, compelling myths, and recognized leaders, such as priests, ministers, and imams.

We cannot say categorically that religion is necessary to altruistic love in humans, but we recognize that the story of the human race through the centuries is that many acts of kindness, ranging from small to extraordinary, have a religious basis. Of course, altruistic and other-regarding behavior is found in nonreligious, secular individuals and groups as well, but the historical record reveals a high degree of correlation between altruistic accomplishments and motivations rooted in a sense of relationship with a Supreme Being who is supremely loving or in a sense of oneness with a Cosmic Love. Actually, reasonable theological explanations of altruistic love among nonbelievers are also available. For example, theistic natural law theory, together with theistic virtue theory, implies that divinely created human nature was intended to engage in loving actions borne out of loving motivations such that it is no surprise that nonbelievers, who are also divinely created, would reflect these aspects of the divine purpose without being committed to their ultimate source. What religious commitment can do, then, is to clarify believers' sense of purpose to love others and strengthen the intention to live accordingly.

The current discussion of whether there can be authentically selfless altruism and altruistic love – which involves theories, findings, insights,

---

[58] Wall, *Mama's Last Hug*, 7; C. Darwin, *The Expression of the Emotions in Man and Animals* (London: Murray, 1872), 132–146.

and principles from biology, psychology, neurology, primatology, philosophy, and religion – is still unsettled. As Stephen Post says, "altruism" among biologists refers simply to actions, not to conscious motivations, that benefit the reproductive success of others at a cost to the self, while "altruistic love" denotes the psychological or motivational aspects of altruism that are of interest to social scientists.[59] On the one hand, biologists and sociobiologists have for many years claimed that either inclusive fitness or reciprocal altruism explains animal behaviors that ensure the actor's own genes are passed on, which makes all apparent altruism a form of underlying egoism. In *On Human Nature*, E. O. Wilson claims that all altruism is "soft core," that is, ultimately selfishly calculated to expect reciprocation, and that it often trades on "pretense and deceit," even on "self-deceit, because the actor is most convincing who believes that his performance is real."[60] In applying this interpretation to the ostensibly noble actions of Mother Teresa, he identifies her egoistic motives: "[I]t should not be forgotten that she is secure in the service of Christ and the knowledge of her Church's immortality."[61] What Wilson calls "hard-core altruism" – or genuine self-sacrifice – is for him not really authentic charity but merely a mindless, irrational death-wish. Evolutionary theorist Michael Ghiselin stated the general conclusion straightforwardly: "[I]f natural selection is both sufficient and true, it is impossible for a genuinely disinterested behavior pattern to evolve."[62]

By contrast, Elliott Sober and David Sloan Wilson advance the opposite perspective – that there is genuine altruism – a view more frequently held among some other philosophers, social psychologists, and anthropologists. Sober and Wilson claim in their book *Unto Others* that unselfish behaviors, without reproductive interests, are widely distributed in biological nature and on full display in human nature – though as one might expect, such observations may have alternative explanations and are somewhat controversial.[63] Sober and Wilson's particular view is based on a form of group

[59] S. G. Post et al., "General Introduction," in S. G. Post et al. (eds.), *Altruism and Altruistic Love*, 4.
[60] E. O. Wilson, *On Human Nature* (Cambridge, MA: Harvard University Press, 1978), 156.
[61] Wilson, *On Human Nature*, 165.
[62] M. T. Ghiselin, "Darwin and Evolutionary Psychology," *Science*, 179, no. 4077 (1973), 967.
[63] See, for example, the case of humpback whales defending other species, which is interpreted by some as genuine altruism and by others as an accidental or inadvertent outcome. E. Stokstad, "Why Did a Humpback Whale Just Save This Seal's Life?," *ScienceMag.org* (August 2016), www.sciencemag.org/news/2016/07/why-did-humpback-whale-just-save-seals-life.

altruism in which acts on behalf of other members of the group go beyond kin selection and reciprocal altruism and affect the larger group. In evolutionary terms, it is entirely plausible, they say, that the survival of the group in competition with other groups takes such forms of group altruism beyond individual reproductive and survival interests. Their conclusion is that "altruism can be removed from the endangered species list in both biology and the social sciences."[64]

In reflecting on how science and religion intersect on the phenomenon of altruism, we can detect familiar philosophical issues in the background: mechanistic versus teleological explanation, reductionism versus a more holistic approach, unrelenting selfishness (even if strategically expressed) versus the capacity for genuine love and self-giving. Frankly, these issues were debated centuries before evolutionary theory was formulated and spawned so many intriguing directions for investigation. And, again, any position on the issues has always been closely related to worldview orientation. For example, naturalist/materialist thinkers have always offered various causal theories, often in categories of the prevailing science, as complete and sufficient explanations of human cognition and behavior, including religious cognition and behavior.

At first blush, it may seem that there is a disagreement between science and religion on the nature of altruism and love because of some conflict between those disciplines. It is true that some theological perspectives insist that humanity was specially created by God and that love is a sign of God's plan for the world. These forms of religion thus completely explain altruism and love and deny that we had an animal prehistory, thereby embracing one form of dualism between spiritual and material aspects of the world that makes science irrelevant to theology. However, it is also true that some scientific viewpoints assert that it is sufficient to explain human morality as purely a product of our evolutionary past, in which case God is unnecessary to explain the phenomenon of love as we know it, thereby tacitly endorsing materialism and reductionism. Each of these two polarized perspectives generates conflict over the nature of altruism and love, but careful analysis of the issues reveals intellectual precommitments that affect how the explanatory roles of science and religion are understood by each extreme.

---

[64] E. Sober and D. S. Wilson, *Unto Others: The Evolution and Psychology of Unselfish Behavior* (Cambridge, MA: Harvard University, 1998), 337.

Moderate positions are often more balanced and more fruitful. In fact, Irving Singer concludes the third volume in his magisterial three-volume study *The Nature of Love* with an encouraging note: "the most promising opportunity for us in the twentieth century is to be found in a synthesis of scientific and humanistic approaches to human affect."[65] In the early decades of the twenty-first century, it is clear that much more science, as well as much more theology, must be put in the mix with the sweep of humanistic disciplines that Singer surveyed on love. Orthodox Christian theology, for example, affirms that God is the ultimate source of love while being fully appreciative of scientific information about how love occurs in the created physical world. Theology can frame the meaning of love in teleological terms, and science can articulate the mechanical processes associated with love. Likewise, a scientific perspective that is not materialist or reductionist about the biological and neurological information regarding altruism can add much to an interdisciplinary discussion that includes insights from theology or philosophy. In any case, the polarities and tensions inherent in the science–religion discussion of altruism also reflect the multiple complex aspects of human nature itself.

[65] I. Singer, *The Nature of Love: The Modern World* (1987; repr. Cambridge, MA: MIT Press, 2009), vol. III, 345.

8 Biology, Ethics, and Debunking Arguments

The application of evolutionary theory to morality has given rise in our day to a field known as "evolutionary ethics." From the beginning, claims about how evolutionary ideas engage ethical thought have sparked controversy. Both Charles Darwin and Herbert Spencer provided their own distinctive evolutionary reasoning about the foundation, content, and function of human morality, while several well-known traditional philosophers, such as Henry Sidgwick, denounced and severely critiqued their views. Given the long-standing connection between ethics and religion, the issues become more complex. In this chapter, we explore the basic issues arising at the intersection of evolutionary ethics and religion – from how areas of ethics are affected to how religion is challenged, with particular attention to theism and Christianity. We also discuss important philosophical work regarding whether evolution debunks realist interpretations of religious ethics and how evolutionary ethics is interpreted by both naturalistic and theistic worldviews.

Biology and Types of Ethical Theory

Although many contemporary ethical philosophers have not recognized the full impact of evolution on their field, the impact is actually quite broad (covering almost every level of ethical theorizing) and very deep (radically reorienting traditional ideas of morality and human agents). To begin engaging the multilevel impact of evolutionary ethics, let us briefly survey standard areas of ethics, the major traditions, and some key issues in relation to evolutionary claims. We start with the most fundamental divide in ethics – between "substantive ethics" (or first-order ethics) and "metaethics" (or second-order ethics). Substantive ethics deals with the content of ethics, the actual principles and recommended behavior for ethical living. Metaethics deals with the nature of ethics, from its ontological ground and epistemic status to the meaning of ethical language and the nature of ethical reasoning.

Substantive ethics, often called normative ethics, is often further divided into "theories of obligation" and "theories of virtue." Leaving the discussion of virtue until later, let us consider the intersection of evolution and the ethics of obligation. Most contemporary thinkers advocating a Darwinian viewpoint stipulate that we all more or less agree on what rules and behaviors count as being moral. Do not lie or defraud. Do not steal. Do not murder. Do not discriminate against or be unkind to people on the basis of race or skin color. Violating such rules is considered immoral by most normal human beings. According to Michael Ruse, most of us agree on the basics of morality, but we should not mistakenly think that incredibly bad societies, such as the Third Reich, deny the basics of everyday morality. Ruse explains,

> What they do is alter the realm of application, arguing that Jews and gypsies and gays are not real people. In other words, as so often, it is not a matter of substantive ethics as such but of empirical claims about realization, applicability, and so forth.[1]

This is not to deny that some individuals and societies actually reject some or all of the content of traditional morality, a point we shall address shortly.

Of course, the familiar basic content of morality closely resembles Christian morality, which has shaped Western society for centuries. Hence, as we pursue the issues, we will put theistic and Christian ideas of morality in interaction with evolutionary ideas. Historically, opinions about this have varied widely. In the nineteenth century, many influential thinkers believed that Darwinian ideas about ethics would lead to an ethic of "survival of the fittest" incompatible with Christian ethics. Besides, if there is no God who gives us morality, these thinkers worried, what is to restrain the savage in all of us? For such reasons, they defended the connection between morality and God against evolutionary thinking in order to maintain human dignity and social order.

---

[1] M. L. Peterson and M. Ruse, *Science, Evolution, and Religion: A Debate about Atheism and Theism* (New York: Oxford University Press, 2016), 176.

Interestingly, Thomas Henry Huxley, a famous nonbeliever and inventor of the term "agnostic," pressed the London School Board for compulsory Bible studies so that children could receive proper moral guidance.[2] Nevertheless, as Huxley argued, if we are "modified monkeys" rather than "modified dirt" (an allusion to the creation imagery of Genesis), then we should follow the scientific evidence wherever it leads us to gain important information about ourselves, including information about our moral behavior.[3] At the level of specifying our substantive duties, then, ethics according to evolutionary thinking might to some degree parallel Christian normative directives – such as "do not steal" and "do not lie and defraud." However, while the duties prescribed by Darwinian and Christian morality might be somewhat similar, the Darwinian naturalist grounds those duties in our biology, not in God, which means the difference is at the level of metaethics. Similarly, the content of Darwinian morality would overlap with, for instance, the content of Kantian ethics, but the Darwinian justification of the common norms would cite biology as the ground, not Kant's rational respect for duty. None of this is to ignore the apparent disagreements between Darwinian ethics on the one hand and traditional and Christian ethics on the other, a topic we take up later. At this point, however, it is clear that ethics must take into account our evolutionary history and the ways biology has shaped our morality.

Darwinian Evolution and the Moral Sense

In 1871 in *The Descent of Man*, Darwin identified morality as "the best and highest distinction between man and the lower animals."[4] Nevertheless, he followed the principles in his previously published *Origin* by suggesting that this capacity is rooted in evolutionary history and, therefore, could accrue to any animal with sufficiently evolved capacities. Conscience, then, according to Darwin, arises when a certain degree of rationality develops and overlays what he calls the "social instincts," interpreting and orchestrating them:

[2] T. H. Huxley, "The School Boards: What They Can Do, and What They May Do," in *Science and Education: Essays by Thomas Henry Huxley* (New York: J. A. Hill and Co., 1904), 210.
[3] Huxley to F. Dyster, January 30, 1859, Huxley Papers, 15.106, Imperial College.
[4] C. Darwin, *The Descent of Man*, 1st edn., 2 vols. (London: Murray, 1871), vol. I, 106.

The following proposition seems to me in a high degree probable – namely, that any animal whatever, endowed with well-marked social instincts, the parental and filial affections being here included, would inevitably acquire a moral sense or conscience, as soon as its intellectual powers had become as well, or nearly as well developed, as in man.[5]

As Robert J. Richards observes, Darwin's full theory of conscience – or the capacity to recognize right and wrong – contains several major elements: a repertoire of social instincts; high enough intelligence and memory to make practical judgments of a complex sort; language to codify behavior and communicate desires, requests, and other information to conspecifics; and habits serving to shape behavior.[6]

A Darwinian narrative account of the origin of conscience or the moral sense begins with the biological fact that humankind is a species of social animal, which lives in a family, group, and society. Social instincts are part of the very "essence" of what a social animal is, instincts formed by natural selection and continually operative in the life of the individual and group. Of course, the social instincts work differently depending on the species – in ants and bees by determining specific roles for individuals in the group; in more complex animals, strengthening tendencies to prefer social life and urging parental care, group cooperation, and altruistic response. According to philosopher of science Soshichi Uchii, however, the "moral sense" or conscience is not the functioning social instincts per se but rather is developed from a combination of the functioning social instincts with sufficiently high intelligence. Social animals that have acquired high intelligence can remember past actions, motives for actions, and feelings associated with past actions – all of which intensify sympathy, which is included in the social instincts. Sympathy – conceived as the ability to "re-present" in one's psychology the feelings of others within oneself – can be extended as the animal acquires better knowledge of others.[7]

---

[5] Darwin, *Descent of Man*, vol. I, 71–72.
[6] R. J. Richards, "Evolutionary Ethics: A Theory of Moral Realism," in M. Ruse and R. J. Richards (eds.), *The Cambridge Handbook of Evolutionary Ethics* (New York: Cambridge University Press, 2017), 143.
[7] S. Uchii, "Darwin on the Evolution of Morality," paper presented at University of Pittsburgh Center for Philosophy of Science International Fellows Conference in Castiglioncello, Italy, May 20–24, 1996.

Nevertheless, strong sympathy for others is still not quite the moral sense, because sympathy may not be stronger on every occasion than motives for sex or food, which can occasion antisocial behavior. So, is what gives rise to conscience the enduring nature of the social instincts? The Darwinian answer is that, when social instincts are frustrated and unsatisfied, a disagreeable feeling remains. In memory, the feelings associated with the social instincts would become dominant. Similarly, agreeable feelings associated with satisfying the social instincts rather than other desires will be pleasantly remembered. This is how distinctively moral feelings arise. Darwin completes the narrative on the "moral sense" by referring to the ability to use some sort of "language," which would communicate the desires of individuals as members of a community. He further suggests that an interplay between sympathy, social norms, and individual moral habits helps make for a "culture" in which common understandings of morality become embedded.

The crucial element for Darwin's account of ethics regards the "social instincts," which are expressions of cooperative and altruistic responses. Natural selection, he contended, could develop responses that seem to benefit the recipient rather than the actor:

> It must not be forgotten that although a high standard of morality gives but a slight or no advantage to each individual man and his children over the other men of the same tribe, yet an enhancement in the standard of morality and increase in the number of well-endowed men will certainly give an immense advantage to one tribe over another. There can be no doubt that a tribe including many members who from possessing in a high degree the spirit of patriotism, fidelity, obedience, courage, and sympathy, were always ready to give aid to each other and to sacrifice themselves for the common good, would be victorious over most other tribes, and this would be natural selection.[8]

Although Darwin scholars debate whether this passage endorses "kin selection" or "group selection" for tribal survival, the basic point is that tendencies toward certain instincts and behaviors were adaptive given the circumstances of our evolutionary path, helping groups survive and in turn improving the well-being of their members. Natural selection has powerfully shaped our evaluative attitudes so that we consider good that which

---

[8] Darwin, *Descent of Man*, vol. I, 166.

promotes the survival of the species. Because of this, E. O. Wilson declared that "the time has come for ethics to be removed temporarily from the hands of the philosophers and biologicized."[9]

In *The Descent of Man*, Darwin sought to determine whether natural selection could account for the observable trait that we call the human moral sense, but he avoided the philosophical function of defining or redefining morality. Instead, he specifically studied human morality "exclusively from the side of natural history" – in accord with principles of that history first exposited in the *Origin*.[10] Group selection, along with a desire to be praised but not blamed by one's community, a natural inclination toward sympathy, and an acquired habit of helping others, rounded out Darwin's account of human morality.[11] While Darwin's account is intentionally descriptive, many evolutionary thinkers, from his own day to ours, construe his ideas to impact the justification of morality and its norms at a deep level, ultimately forcing a reinterpretation of ethics. For many, Darwin's account has serious implications for belief in objective morality and the human ability to know moral truth, matters that are important to classical philosophical and religious ethics.

The Darwinian account of the social instincts was further seen as having implications for both classical and religious conceptions of human uniqueness. For example, after reading *Origin*, Christian minister Leonard Jenyns wrote to Darwin:

> One great difficulty to my mind in the way of your theory is the fact of the existence of Man [that you seem to suppose] is to be considered a modified and no doubt greatly improved orang! ... Neither can I easily bring myself to the idea that man's reasoning faculties and above all his moral sense could ever have been obtained from irrational progenitors, by mere natural selection – acting however gradually and for whatever length of time that may be required. This seems to me doing away altogether with the Divine Image that forms the insurmountable distinction between man and brutes.[12]

Since Christianity – as well as the other great monotheisms, Judaism and Islam – typically cites the moral sense or conscience as an important way in

---

[9] E. O. Wilson, *Sociobiology*, 25th anniversary edn. (Cambridge, MA: Harvard University Press, 2000), 562.

[10] Darwin, *Descent of Man*, vol. I, 71.   [11] Darwin, *Descent of Man*, vol. I, 158–184.

[12] L. Jenyns to Darwin, January 4, 1860, Letter no. 2637A, Darwin Correspondence Project, www.darwinproject.ac.uk/letter/DCP-LETT-2637A.xml.

which humanity is special in creation, the Darwinian view that our morality has a primate prehistory is a major concern.

Evolution and the Institution of Morality

The Darwinian natural history approach starts with the struggle for existence: "[E]ach organic being is striving to increase at a geometrical ratio; that each at some period of its life, during some season of the year, during each generation or at intervals, has to struggle for life, and to suffer great destruction."[13] Most evolutionary theorists, then, say that, given natural selection, a morality based on sympathy was useful for survival in our evolutionary past.[14] Yet, we ask, how in biological terms did we move from a "war of all against all" – *bellum omnium contra omnes*, as Thomas Hobbes calls it – to cooperative and altruistic behavior? Darwin's answer begins with the observation that "each man would soon learn from experience that if he aided his fellow-men, he would commonly receive aid in return." He states that this "low motive" still fosters the habit of aiding one's fellows and that the habit strengthens sympathetic response prompting benevolent actions. The habits of sympathetic response then become heritable down through many generations.[15] Here we have the outline of a Darwinian genealogy of morals.

Wilson argues that altruistic behaviors, when biologicized, turn out to be forms of egoism, expressions of self-interest.[16] Biologist Richard Dawkins claims that "selfish genes" seeking their own replication are the ultimate causes of human moral behavior.[17] Darwin knew that the biological origin of morality requires inheritance, which we now know is accomplished by the gene. Nonetheless, the Darwinian scenario is clear: in the contingencies of the evolutionary landscape, certain behaviors are adaptive; propensities for such behaviors are also adaptive; insofar as these propensities are genetic, they are heritable. Thus, the "social instincts" have passed from our animal ancestors to form the emotional basis of human moral behavior: "[T]hose communities, which included the greatest number of the most sympathetic members, would flourish best and rear the greatest number of offspring."[18]

---

[13] C. Darwin, *On the Origin of Species* (London: Murray, 1859), 78–79.
[14] For example, see R. Joyce, *The Evolution of Morality* (Cambridge, MA: MIT Press, 2006).
[15] Darwin, *Descent of Man*, vol. I, 163–164.   [16] Wilson, *Sociobiology*, 562–564.
[17] R. Dawkins, *The Selfish Gene* (New York: Oxford University Press, 1976), 18–19.
[18] Darwin, *Descent of Man*, vol. I, 82.

Darwin, of course, added to his theory of sympathy a theory of the desire to receive praise and avoid blame:

> But there is another and much more powerful stimulus to the development of the social virtues, namely, the praise and the blame of our fellow-men. The love of approbation and the dread of infamy, as well as the bestowal of praise or blame, are primarily due ... to the instinct of sympathy; and this instinct no doubt was originally acquired, like all the other social instincts, through natural selection.[19]

Of course, enforcement and punishment are not unknown in other species, but our concern is with their function in human morality.

Evolutionary theorists William Harms and Brian Skyrms maintain that certain evolved social behaviors became normative in human society due to enforcement mechanisms. Typically, society reacts strongly to noncooperators with enforcement behaviors such as disapproval, punishment, and ostracism, even when at a cost to the enforcers. The felt "normative force" of social rules, argue Harms and Skyrms, is due to the severity of the negative enforcement. Thus, the normative force of moral statements ("murder is wrong," "stealing is wrong") is that of an enforcement signal, which gets internalized by members of the group who then believe the statements are "true" and "ought" to be followed. Harms and Skyrms say that their ideas move a long way toward "a materialist theory of norms which avoids the pitfalls of relativism" because they are rooted in objectively real but evolved social reactions.[20]

The obvious question is whether a materialist or naturalist perspective can provide a viable objectivist account of what is typically meant by the normative force of ethics. The kind of objectivity in ethics that Harms and Skyrms accept anchors the normative character of ethics in the strength of reactions in human groups, which are ultimately anchored in biological fitness. Yet religious believers are typically suspicious of this kind of move because they trace, in one way or another, the objective normative force of ethics to a transcendent divine source, a deity that is worthy of worship and obedience. John Henry (Cardinal) Newman made the point: "If, as is the case, we feel responsibility, are ashamed, are frightened, at transgressing the voice

---

[19] Darwin, *Descent of Man*, vol. I, 164.
[20] W. Harms and B. Skyrms, "Evolution of Moral Norms," in Ruse and Richards (eds.), *The Oxford Handbook of Philosophy of Biology* (New York: Oxford University Press, 2008), 446.

of conscience, this implies that there is one to whom we are responsible, before whom we are ashamed, whose claims upon us we fear."[21] For Newman, moral principles stem from the God of Christianity: an intelligent, moral, personal being who gives us conscience as his guidance in us:

> These feelings in us are such as require for their exciting cause an intelligent being; we are not affectionate towards a stone; we do not feel shame before a horse or dog; we have no remorse or compunction on breaking merely human law; yet so it is, conscience excites all these painful emotions: confusion, foreboding, self-condemnation; and on the other hand it sheds upon us a deep peace, a sense of security, a resignation and a hope, which there is no sensible, no earthly, object to elicit.[22]

For Newman, moral principles stem from the God of Christianity, an intelligent, moral, personal being who gives us conscience to understand the principles.

The religious worry is that, if conscience can produce all the moral emotions, and if the emotions can be explained as products of biological factors, then a major reason for believing in God as the author of human moral life has been diminished. In *Descent*, Darwin theorizes as follows:

> It has, I think, now been shewn that man and the higher animals, especially the Primates, have some few instincts in common. All have the same senses, intuitions, and sensations, – similar passions, affections, and emotions, even the more complex ones, such as jealousy, suspicion, emulation, gratitude, and magnanimity; they practice deceit and are revengeful; they are sometimes susceptible to ridicule, and even have a sense of humor; they feel wonder and curiosity; they possess the same faculties of imitation, attention, deliberation, choice, memory, imagination, the association of ideas, and reason, though in very different degrees. The individuals of the same species graduate in intellect from absolute imbecility to high excellence. They are also liable to insanity, though far less often than in the case of man.[23]

The similarity of human and animal emotions (particularly primate emotions) raises the question of whether morality can be maintained as objective and tied to a transcendent source; or is it completely grounded in our biology?

[21] J. H. Newman, *A Grammar of Assent* (New York: Catholic Publishing Society, 1870), 109.
[22] Newman, *Grammar of Assent*, 110.
[23] C. Darwin, *The Descent of Man*, 2nd ed. (London: John Murray, 1874), 79.

Religion or Biology as the Basis of Morality?

Evolutionary theory presents various challenges to religious understandings of morality. One challenge is to the moral argument, which is important to natural theology or apologetics. The great medieval philosopher Thomas Aquinas argued that the moral law in human life – the "natural law" – is a concrete reflection of divine eternal law. Immanuel Kant argued that our conviction that morality is absolute and universal makes rational sense only if we postulate that there is a God (short of being able to provide a discursive argument for God). In the middle of the twentieth century, Christian thinker C. S. Lewis presented his own version of the moral argument: that morality seems to be an important way that "the power behind the universe" is communicating to us how he wants us to act. All such moral arguments are put under pressure by evolutionary thinking.

Another challenge is that an evolutionary account can be advanced as a total and sufficient explanation of morality – one linked to an alternative vision of the world in which humans are not special creatures. Empirical support for an evolutionary theory of morality seems abundant, given the observational and experimental data in studies of other eusocial species (species with the highest degree of sociality). Familiar studies of altruism and cooperation in species such as ants and prairie dogs can now be readily augmented by several decades of research on monkey, chimpanzee, and ape societies. For example, chimpanzees, which cannot swim, have drowned in zoo moats trying to save others. We even know that, if rhesus monkeys are given an opportunity to acquire food by pulling a chain that triggers an electric shock to a close companion rhesus monkey, these monkeys will refuse to pull the chain and thus starve themselves for a significant time.[24]

Primatologist Frans de Waal has mounted extensive evidence that primates constrain their behavior for the benefit of the group. Given the closeness between the higher primates and humans, these constraints must be viewed as part of the human inheritance that forms the basis of emotional responses and behaviors around which morality has emerged. In the 1960s, de Waal's work on aggression in nonhuman primates revealed that it was common in chimpanzees for there to be consolation of the loser in combat,

---

[24] S. Wechkin, J. H. Masserman, and W. Terris Jr., "Shock to a Conspecific as an Aversive Stimulus," *Psychonomic Science*, 1, no. 2. (1964), 47–48.

which is evidence of a level of empathy that only apes and humans seem to possess. A key function of empathy in primates, particularly in chimpanzees, seems to be aimed at ending hostilities – apparently, for the good of the group more than for individual relationships.[25] In later experiments, de Waal with psychologist Sarah Brosnan found that capuchin monkeys had a strong aversion to receiving unequal rewards for performing the same task (handing a rock to the experimenter). When receiving the same wage – a slice of cucumber – for the task, two female capuchin monkeys, in side-by-side separate cages, were perfectly willing to repeat the task an indefinite number of times. However, when one monkey received a grape (a much better food) for the same task, the monkey receiving only a cucumber became upset and so agitated that it threw the cucumber back at the experimenter.[26]

The experiment also revealed the perception of unfairness by the monkey receiving the better reward, again showing a higher level of cognition seen only in apes and humans. Making this point, de Waal states, "The chimpanzees may refuse a grape if the other one doesn't also get a grape. So they want to equalize the outcome."[27] Similar experiments have been repeated on crows and dogs, with similar aversion reactions to social inequity. In their *Nature* article entitled "Monkeys Reject Unequal Pay," Brosnan and de Waal theorize that "during the evolution of cooperation it may have become critical for individuals to compare their own efforts and pay-offs with those of others."[28] For de Waal, empathy – as the ability to learn and follow social rules, and to engage in reciprocity and peacemaking – forms the basis of sociality. Human morality, he concludes, has grown out of basic primate sociality, but with two additional levels of sophistication unparalleled among animals: stronger enforcement of the moral codes of their societies that utilize reward, punishment, and reputation building as well as employment of complex rational judgment.

---

[25] Typically, a female will bring the winner and loser males together – seeming to end discord and maximize cooperation in the group. Stated by F. de Waal in N. Wade, "Scientist Finds the Beginnings of Morality in Primate Behavior," *New York Times*, March 20, 2007, www.nytimes.com/2007/03/20/science/20moral.html.

[26] See the actual video: https://youtu.be/meiU6TxysCg.

[27] F. de Waal interviewed by S. Strockes, "Why Monkeys Care about Fairness and What It Means for Us," City Café 90.1 FM WABE, November 12, 2014, www.wabe.org/why-monkeys-care-about-fairness-and-what-it-means-us/.

[28] S. F. Bronsnan and F. de Waal, "Monkeys Reject Unequal Pay," *Nature* 425, no. 6955 (2003), 297.

Because of these scientific findings, many suggest that our biology is the "ground" of morality, in forming both the capacity to think morally and the content of morality. Moral philosopher Justin Horn states,

[J]ust as the pressure of natural selection played a significant role in bringing it about that tigers have sharp claws (rather than dull ones) and that zebras are speedy (rather than slow), so too did the pressure of natural selection play a significant role in bringing it about that human beings have a very strong tendency to regard certain things as morally valuable or morally reprehensible. Examples of evolutionarily favored moral attitudes might include the widespread positive moral regard enjoyed by activities such as caring for one's own children and reciprocating benefits provided by others, and the negative moral regard commonly held for defecting from agreements or casually harming one's kin.[29]

Of course, all of the impressive evidence still requires philosophical interpretation in order to determine its ultimate bearing on religion. Some hold that God and evolution are dichotomous, such that evolutionary morality preempts morality anchored in the God of religion. Others hold that God, as the source of morality, employed evolutionary processes to mediate morality in human life.

Epistemological Debunking Arguments

The claim of evolutionary ethics – that human morality is rooted in our emotions, which are shaped by our evolutionary history – puts pressure on both classical and Christian ethical theories that treat morality primarily as a matter of reason. Traditional views that link ethics to reason also typically maintain that ethical knowledge is objective and absolute, whereas views that classify ethics as a matter of emotion are usually associated with claims that ethics is subjective and relative. Evolutionary theory reinvigorates the debate over the relative plausibility of these two broad views.

David Hume, in the eighteenth century, asserted that morality is a matter of the "sentiments" or "passions," or, as we might say, "emotions" or "feelings." He famously set his nonobjectivist viewpoint in opposition to the thinking of most moral philosophers, ancient and modern, who hold that

[29] J. Horn, "Evolution and the Epistemological Challenge to Moral Realism," in Ruse and Richards (eds.), Handbook of Evolutionary Ethics, 115.

reason governs the passions, when he declared that "reason is and ought only to be the slave of the passions."[30] As Hume explained, the rational apprehension of an ethical principle or law is powerless to move us to action, for the motivational engine of human actions is the passions. Hume shrewdly observes the differential sense of obligation between relatives and nonrelatives based on emotional closeness:

> A man naturally loves his children better than his nephews, his nephews better than his cousins, his cousins better than strangers, where everything else is equal. Hence arise our common measures of duty, in preferring the one to the other. Our sense of duty always follows the common and natural course of our passions.[31]

The discussion of kin selection in Chapter 7 makes Hume seem prescient about this much-studied topic in contemporary evolutionary biology.

The recognition by Darwin and contemporary evolutionary biologists that our "passions" or "social instincts" are shaped by natural selection is typically taken to suggest that ethics is not only subjectively based but also relative in the sense of not being absolute and universal. Moral instincts shaped by the vicissitudes of our evolutionary path might have been different had the contingencies along the way been different, as Darwin stated:

> If ... men were reared under precisely the same conditions as hive-bees, there can hardly be a doubt that our unmarried females would, like the worker-bees, think it a sacred duty to kill their brothers, and mothers would strive to kill their fertile daughters, and no one would think of interfering.[32]

Justin Horn states the counterfactual at work here: "Had we evolved differently, we would make moral judgments with very different content."[33] Traditional moral theories typically assume that ethical judgments about right and wrong are absolute or unchanging because they are anchored either in unchanging human nature or in the will of deity.

The evolutionary information is prelude to the debate among contemporary analytic philosophers over whether a realist ethical position – asserting the objectivity and universality of moral principles – is viable in light of an evolutionary conception of the world, which can be construed as antirealist.

---

[30] D. Hume, *A Treatise of Human Nature* (1739–1740), 2.3.3.4.
[31] Hume, *A Treatise of Human Nature*, 3.2.1.18.   [32] Darwin, *Descent of Man*, vol. I, 73.
[33] Horn, "Evolution and the Epistemological Challenge to Moral Realism," 115.

The metaethical debate further divides into epistemological disagreements about the justification of moral beliefs and ontological disagreements about the status and nature of moral truth.

In the mid-1980s, E. O. Wilson and Michael Ruse identified the epistemological difficulty that evolution creates for ethics:

> [W]hat Darwinian evolutionary theory shows is that this sense of "right" and the corresponding sense of "wrong," feelings we take to be above individual desire and in some fashion outside biology, are in fact brought about by ultimate biological processes.[34]

Citing the essentially causal explanation for ethics provided by biology, moral philosopher Richard Joyce explicitly formulated the case for epistemological metaethical antirealism in his 2006 book *The Evolution of Morality*, arguing that plausible evolutionary reasons for thinking that the "moral sense" is a biological adaptation have a debunking effect on moral beliefs. For Joyce, "our moral beliefs are products of a process that is entirely independent of their truth."[35] The clear antirealist implication is that "we have no grounds one way or the other for maintaining these beliefs."[36] In this light, the very enterprise of ethics is without justification; moral beliefs cannot constitute knowledge.

Philosopher Sharon Street clearly pinpoints the antirealist challenge to the realist:

> [T]he realist must hold that an astonishing coincidence took place – claiming that as a matter of *sheer luck*, evolutionary pressures affected our evaluative attitudes in such a way that they *just happened* to land on or near the true normative views among all the conceptually possible ones.[37]

According to Street and other antirealists, the evolutionary information supplies the basis for a "debunking argument" – an undercutting epistemic defeater for our moral beliefs – which we sketch as follows:

1. Evolutionary forces shaped our moral capacities and/or beliefs.

---

[34] M. Ruse and E. O. Wilson, "Moral Philosophy as Applied Science," *Philosophy*, 61 (1986), 179.

[35] Joyce, *Evolution of Morality*, 211.   [36] Joyce, *Evolution of Morality*, 211.

[37] S. Street, "Reply to Copp: Naturalism, Normativity, and the Varieties of Realism Worth Worrying About," *Philosophical Issues*, 18 (2008), 208–209; emphasis added.

2. Evolutionary forces are concerned with adaptive fitness, not with moral truth.

3. It is highly unlikely that our moral beliefs are in touch with anything that might be called "moral truth."

Therefore,

4. Our moral beliefs do not constitute knowledge because they are unjustified.

So, even if there are objective moral truths, our moral beliefs are "independent of" – and "insensitive to" – these truths because our moral beliefs are fitness driven. Thus, moral beliefs have no rational justification and cannot constitute knowledge, making moral skepticism the ultimate result.

In order to avoid moral skepticism, a number of philosophers, some naturalists and some theists, have replied to evolutionary debunking arguments along several lines. One line of response is that the debunking argument appears to be a non sequitur in the form of an elementary genetic fallacy – that is, the mistake of thinking that an explanation of the origin of an idea is automatically a reason to discredit that idea. Nicholas Sturgeon, a nontheistic moral philosopher, argues that just because there are reasons to suspect independence of moral beliefs from moral truths, we cannot legitimately infer that there are no reasons whatsoever for thinking a relevant dependence relation obtains between the causes of our moral beliefs and what would make them true. After all, he observes, "Many moral explanations appear to be good explanations ... that are not obviously undermined by anything else we know," including evolutionary science.[38] He adds, "Sober people frequently offer such explanations of moral observations and beliefs."

To counter such moves, Wilson and Ruse assert that "ethics as we understand it is an illusion fobbed off on us by our genes in order to get us to cooperate."[39] The sociobiological story is familiar: the pressures of natural selection exerted enormous influence on human psychology, including the hardwiring of "epigenetic rules," which Wilson and Ruse take to be widely distributed propensities to believe and behave in certain ways that developed

---

[38] N. Sturgeon, "Moral Explanations," in G. Sayre-McCord (ed.), *Essays on Moral Realism* (Ithaca, NY: Cornell University Press, 1988), 239.

[39] M. Ruse and E. O. Wilson, "The Evolution of Morality," *New Scientist*, 17 (1989), 51.

through the interaction of human genetics and human culture. Ruse explains,

> The Darwinian argues that morality simply does not work (from a biological perspective), unless we believe that it is objective. Darwinian theory shows that, in fact, morality is a function of (subjective) feelings; but it shows also that we have (and must have) the illusion of objectivity.[40]

Thus, we widely think our subjective sentiments are more than just sentiments – that they are objective – even though they are not.

For Ruse, then, the sense of "oughtness" we feel about moral rules – what we might call their normative force – is epigenetically determined and does not reflect genuine epistemic contact with moral truth. Ruse often makes the point by drawing from Hume's famous "is/ought" problem: that statements of obligation cannot be logically derived from statements of fact. In other words, from the fact that it "is" the case that our conditioned moral sentiments prefer certain actions, we cannot deduce that we "ought" to prefer those actions in some overarching sense. Ruse is a good example of an evolutionary naturalist who holds an antirealist and subjectivist interpretation of ethics and denies that morality has a foundation in the realist sense. There are a few naturalists who accept the evolutionary account of ethics and yet try to offer interpretations of ethics that are meant to be realist and objectivist, but we cannot pursue these here.[41]

Theistic philosophers generally have always defended against various sorts of skepticism about our cognitive faculties, including the moral sense. In Hume's own day, theistic philosopher Thomas Reid agreed with Hume that there are beliefs and principles "which the constitution of our nature leads us to believe, and which we are under a necessity to take for granted in the common concerns of life."[42] However, for Reid, being produced by our constitution confers positive epistemic status and does not occasion the skepticism of Hume. According to Reid's epistemological realism, these beliefs are part of the "common sense" of the human race – beliefs that all

---

[40] M. Ruse, *Taking Darwin Seriously: A Naturalistic Approach to Philosophy* (Amherst, NY: Prometheus Books, 1998), 253.

[41] For example, see R. Shafer-Landau, "Moral Realism and Evolutionary Debunking Arguments" and W. J. FitzPatrick, "Why Darwinism Does Not Debunk Objective Morality," in Ruse and Richards (eds.), *Handbook of Evolutionary Ethics*, 175–201.

[42] T. Reid, *An Inquiry into the Human Mind on the Principles of Common Sense*, D. R. Brooks (ed.) (1785; repr. Edinburgh: Edinburgh University Press, 1997), 33.

normal human beings take to be true at such a deep level that there is nothing deeper, no premise more fundamental, from which they can be inferentially derived or argumentatively refuted.

Theistic philosopher Alvin Plantinga has helped revive discussions of realism in contemporary epistemology, often relying on insights from Reid. Plantinga's theory is that "basic beliefs," which arise immediately and without inference from faculties such as memory and perception, can count as knowledge. He argues that a basic belief is "rationally warranted" if and only if it is the product of a belief-producing mechanism that is truth-aimed and functioning properly within the environment for which it was designed.[43] This account accommodates those perceptual, memory, testimonial, and even metaphysical beliefs that are the guides of common life – and are among the fund of common native beliefs with which we begin in constructing and assessing any of our theories.

For our present interest, Plantinga does not deny the evolutionary conditioning of our moral sense. Yet he contends that the human moral sense produces reliable "moral beliefs" that count as knowledge – on the condition that the moral faculty is truth-aimed. He argues that the evolutionary account must be interpreted within some overall worldview in order to decide this question. In Street's naturalist worldview, for instance, the probability that evolutionary adaptiveness and moral truth coalesce is extremely low, making any correlation a fluke or pure luck. However, as both Plantinga and Reid maintain, in a theistic worldview that accepts the evolutionary process, the odds would be high that adaptiveness and truth would coalesce, arguably close to 100 percent. In this context, a scientific explanation of morality in evolutionary terms augments rather than undermines our confidence in our moral sense and our received moral knowledge. Reid states the following: "That conscience which is in every man's breast, is the law of God written in his heart, which he cannot disobey without acting unnaturally, and being self-condemned."[44] Although evolution fuels continuing realist/antirealist epistemological debates, evolution is also a source of controversy over the ontological status of morality.

---

[43] A. Plantinga, *Warrant and Proper Function* (New York: Oxford University Press, 1993).
[44] T. Reid, *Essays on the Active Powers the Human Mind* (Cambridge, MA: MIT Press, 1969), 365.

Ontological Debunking Arguments

It is not unusual for epistemological antirealists about ethics to hold onto-logical antirealism as well. Previously, we saw that Wilson and Ruse assert that ethics is an illusion, which is tantamount to the ontological claim that there are no moral facts. Philosophical naturalists generally assert that the world contains only natural properties and facts and thus offer explanations of why humans mistakenly believe that there are moral properties and facts. As Joyce points out, the view that moral beliefs are actually false is even stronger than the epistemological position that they are without warrant.[45]

In 1977, J. L. Mackie stated that "there exist no objective values," thereby giving birth to what we call modern error theory. Consider his expansion of the point:

> [T]hat values are not objective, are not part of the fabric of the world, is meant to include not only moral goodness, which might be most naturally equated with moral value, but also other things that could be more loosely called moral values or disvalues – rightness and wrongness, duty, obligation, an action's being rotten and contemptible, and so on.[46]

If there were objective values, Mackie continues, "they would be entities or qualities or relations of a very strange sort, utterly different from anything else in the universe."[47] Objective values would be "metaphysically queer" – and they could be known only by some "special faculty of moral perception or intuition" that is itself weird and unlike any other mode of knowledge.

Joyce, Street, Ruse, and others who advance epistemological debunking arguments incorporate evolutionary explanations of morality that deny ontologically that there are moral properties and facts. Thus, they too follow in the tradition of error theory and create their own versions of ontological debunking arguments.[48] Ruse, for example, combines his claim that evolu-tion has produced certain useful subjective responses that are the ground of morality with the additional claim that there actually are no ontologically objective moral facts about which our beliefs could be true anyway. Ruse argues that evolutionary explanation shows that there is no *"independent,*

---

[45] Joyce, *Evolution of Morality*, 180.
[46] J. Mackie, *Ethics* (Harmondsworth: Penguin, 1977), 15.   [47] Mackie, *Ethics*, 38.
[48] Joyce, *Evolution of Morality*; S. Street, "A Darwinian Dilemma for Realist Theories of Value," *Philosophical Studies*, 127, no. 1 (2006), 109–166; Ruse, *Taking Darwin Seriously*.

objective, moral code – a code which ultimately is unchanging and not dependent on the contingencies of human nature."[49] The absence of moral properties or facts means that there are no objective moral truths to be epistemically acquired. Thus, our moral judgments, construed in realist terms as being independent in important ways from our attitudes or stance, are false in toto. In summarizing the contemporary literature offering evolutionary accounts of the development of the human moral sense, Joyce approvingly states, "It was no background assumption of that explanation that any actual moral rightness or wrongness existed in the ancestral environment."[50]

Mark Linville, a theistic moral philosopher, has responded to distinctively ontological debunking arguments. He points out that, if it is an assumption that there are no moral properties or facts, then philosophical naturalism has been imported into the discussion without argument, which begs the question at issue. Additionally, argues Linville, if it is simply assumed that morality is a product of or reducible to natural properties and facts, we get various conceptual oddities, such as the following:

> If no moral properties exist, such as depravity, then Hitler was not actually depraved, but we would still believe that he was.[51]

Philosopher Simon Blackburn ostensibly embraces the oddity, commenting that the naturalist "asks no more than this: a natural world, and patterns of reaction to it."[52] Assuming that the evolution of our moral capacity remains the same, Linville rejects as implausible the idea that, in a world with no moral properties, we would react morally in the same way to Hitler's natural properties.

Linville, Plantinga, and many other theists who accept science offer alternative visions of the evolution of morality. They generally maintain that God created the universe to unfold through evolutionary processes that eventually brought rational–social–relational creatures into being. What we shall call "evolutionary theism" or "theistic evolutionism" entails a general genealogy of morals competitive with that of evolutionary naturalism.

---

[49] M. Ruse, *The Darwinian Paradigm* (London: Routledge, 1989), 269; emphasis added.
[50] Joyce, *Evolution of Morality*, 183.
[51] Adapted from M. Linville, "The Moral Argument," in W. L. Craig and J. P. Moreland (eds.), *The Blackwell Companion to Natural Theology* (Malden, MA: Wiley-Blackwell, 2012), 405.
[52] S. Blackburn, *Spreading the Word* (Oxford: Clarendon Press, 1984), 182.

The theistic narrative is that God somehow guided natural processes within creation to achieve various higher purposes, including the development of reliable moral capacities that recognize objective moral properties and facts. Christian philosopher Robert Adams states,

> If we suppose that God directly or indirectly causes human beings to regard as excellent approximately those things that are Godlike in the relevant way, it follows that there is a causal and explanatory connection between facts of excellence and beliefs that we may regard as justified about excellence, and hence it is in general no accident that such beliefs are correct when they are.[53]

Thus, these theists claim to identify an epistemic "tracking relation": we have the basic moral beliefs we do because our moral faculty is reliably aimed at truth and there are moral truths to be epistemically acquired. Generally speaking, they argue, moral beliefs possess reproductive fitness because they are true; fitness is not the only reason we have them. Theists might acknowledge that the adaptive value of a human father caring for his children is unquestionable on pragmatic grounds, but they make the further claim that it is also objectively true that fathers ought to care for their children regardless of the adaptive value of doing so.

The difference in worldview understandings of how evolution affects ethics could not be sharper. Wilson refers to a sharp distinction between "transcendentalist" ethical theories that posit objective "moral guidelines outside the human mind" and "empiricist" ethical theories that posit that "moral values come from humans alone; God is a separate issue."[54] Taking Hume, Darwin, and contemporary sociobiologists to be on his empiricist side, Wilson places on the transcendentalist side all divine command theorists emphasizing "God's will," Kantian thinkers advocating the "categorical imperative," and Thomistic philosophers relying on natural law.[55]

As a brief case study, let us take Thomism, which entails that Wilson holds a false dichotomy because a natural law framework allows ethics to be grounded both in a scientific account of human nature, including an accurate sociobiological account, and in a realist, objectivist understanding of that status of morality. Aquinas argues that natural law morality is "participation in the eternal law on the part of the rational creatures."[56] Ruse has

---

[53] R. Adams, *Finite and Infinite Goods* (New York: Oxford University Press, 1999), 70.
[54] E. O. Wilson, *Consilience: The Unity of Knowledge* (New York: Knopf, 1998), 260–261.
[55] Wilson, *Consilience*, 261.   [56] Aquinas, *Summa Theologiæ*, 2b.94.2.

commented that, since morality must be shared by all, a natural law position could look initially plausible.[57] On the other hand, Ruse has also stated that the relativistic character of evolutionary ethics is incompatible with the absolute character of any ethic purporting to be Christian.[58] A standard Thomistic reply to Ruse recognizes both the absolute and the relative aspects of morality – that moral principles are anchored in universal objective human nature but are expressed somewhat differently across cultures and religions and thus subject to situational application. In one sense, the Thomist can agree with Hume's is/ought barrier – that moral values and obligations cannot be deduced from empirical facts – while still endorsing the idea that there is a metaphysical fact of human nature in which all values and obligations are grounded.

[57] M. Ruse, *Can a Darwinian Be a Christian?* (New York: Cambridge University Press, 2001), 203.

[58] Peterson and Ruse, *Science, Evolution, and Religion*, 182, also see 193.

9   Biological Accounts of Religion

Historically, many thinkers have assumed that religion has a natural origin, but today researchers in the biosciences typically assume that its natural origin falls within the scope of evolutionary theory. For all scientific theories of religion, the general aim has always been to identify empirically the causal processes that operate in all human behaviors that are manifested in the phenomenon of religion. In our day, sociobiologists, evolutionary psychologists, and neuroscientists have advanced theories about the evolutionary causes of religious cognition and behavior. Although the life and behavioral sciences are becoming increasingly unified under the overarching paradigm of evolutionary theory, a consensus on the evolutionary explanation of religion – its development and function – may not quite yet be on the horizon.

In the present chapter, we initially review historical efforts to provide purely natural accounts of religion and then turn to contemporary evolutionary accounts. The discussion these days includes three broad types of evolutionary explanation of religion, which we label straightforwardly: adaptive, nonadaptive, maladaptive. As we develop these theories, we consider various religious responses. At the end, we offer a philosophical appraisal of the current state of the discussion.

The Scientific Study of Religion

Daniel Dennett has claimed that scientists and researchers "are now beginning, for the first time, to study the natural phenomenon of religion through the eyes of contemporary science."[1] Although we are not witnessing literally

[1] D. Dennett, *Breaking the Spell: Religion as a Natural Phenomenon* (New York: Viking Penguin, 2006), 31.

the "first time" that religion has been studied as a natural phenomenon, we are seeing empirical scientists working out of credible theoretical frameworks to develop more rigorous, testable theories in an effort to explain religion causally. The eighteenth-century skeptic David Hume expressed interest in natural causes in his *Natural History of Religion*, which seeks to address two key questions "concerning its foundation in reason, and ... origin in human nature."[2] However, earnest inquiry into the bases of faith – both rational argumentation for religious belief and the innate dispositions for faith – hardly originated with Hume. Such inquiry enjoys a centuries-long tradition within the believing community itself, from Augustine through Anselm and Aquinas to the present day.

In any case, contemporary science pursues the natural causes of religion, Hume's second concern, producing a number of interesting theories that, if true, may have implications for the rationality and truth of religious belief, Hume's first concern. In *Dialogues Concerning Natural Religion*, Hume himself provided what he thought were effective rebuttals of basic theistic arguments for the reasonableness of belief in God and concluded that theistic religion has no credible rational foundation. Together with his natural history of religion, which attributes the emergence of religion to our animal instincts such as self-preservation, Hume provides an explanation for why religion persists even when it lacks what he would consider rational justification for its beliefs. In many ways, the current intellectual climate reflects Hume's assessment. Although scientists operating by methodological naturalism are supposed to be metaphysically neutral, scientists who also believe that religious beliefs lack adequate rational justification are liable to think that natural explanations are the only kind of viable explanation. Thus, there can be a strong tendency to construe the empirical study of religion in terms of metaphysical naturalism, which tacitly combines causal explanations with a philosophical worldview.

Atheist philosopher Kai Nielsen makes the point in his *Naturalism and Religion*:

> [T]here are no sound reasons for religious belief: there is no reasonable possibility of establishing religious beliefs to be true; there is no such thing as religious knowledge or sound religious belief. But, when there are no good

---

[2] D. Hume, introduction to *The Natural History of Religion* (1757, 1777).

reasons for religious belief and [this is] tolerably plain to informed and impartial persons not crippled by ideology or neurosis, and yet religious belief . . . persists in our cultural life, then it is time to look for the *causes* of religious beliefs: causes which are not also reasons justifying religious belief . . . . Here questions about the origin and functions of religion become central.[3]

A large portion of the intellectual community agrees that causal explanations are sufficient to explain religion and religious beliefs. Both classical and contemporary efforts to provide rational grounds for religious beliefs are, therefore, irrelevant to the way "explaining" religion has come to be understood.

Biologist Jerry Coyne rejects supernatural explanation for the dual reasons that he is a committed naturalist and that proposals about the supernatural in science are unfruitful:

We don't reject the supernatural merely because we have an overweening philosophical commitment to materialism; we reject it because entertaining the supernatural has never helped us understand the natural world.[4]

Biologist Richard Lewontin echoes the sentiment, claiming that scientists "have a prior commitment, a commitment to materialism," such that we must construct "an apparatus of investigation and a set of concepts that produce material explanations."[5] We remember Alvin Plantinga's worries from Chapter 3 that intellectual fairness suffers if scientists are allowed naturalistic background philosophical assumptions but not supernaturalistic assumptions.

Christian biologist Jeffrey Schloss comments that, even setting aside the issues regarding demarcation between science and nonscience, and regarding the worldview assumptions that influence a scientist's interpretation of scientific findings, we might try to arrive at a broadly shared methodological approach to the scientific study of religion:

That approach simply involves not invoking the supernatural; it can be characterized as not using the existence of the objects of religious beliefs in explanations of those beliefs.[6]

[3] K. Nielsen, *Naturalism and Religion* (Amherst, NY: Prometheus Books, 2001), 35.
[4] J. A. Coyne, "Don't Know Much Biology," Edge, June 5, 2007, www.edge.org/conversation/jerry_coyne-dont-know-much-biology.
[5] R. Lewontin, "Billions and Billions of Demons," *New York Review of Books*, 44 no. 1 (2001), 31.
[6] J. P. Schloss, "Science Unfettered or Naturalism Run Wild?," in J. Schoss and M. J. Murray (eds.), *The Believing Primate* (New York: Oxford University Press, 2010), 9.

We still need at least a general characterization of religion such that it can be the object of scientific investigation, although an agreed-upon definition of religion has been lacking in scientific accounts.

In Chapter 1, we offered this broad definition: "[R]eligion is constituted by a set of beliefs, actions, and experiences, both personal and collective, organized around a concept of an ultimate reality that inspires or requires devotion, worship, or a focused life orientation." This ultimate reality may be understood as a unity or a plurality, personal or nonpersonal, divine or not, differing from religion to religion. Dennett's proposed definition is that religions are "social systems whose participants avow belief in a supernatural agent or agents whose approval is to be sought."[7] Although religion is more complex than Dennett allows, with many different expressions across social and ecological contexts, we develop our discussion along the lines he suggests. Our approach allows the scientific study of various religious data – beliefs (propositional truth-claims), experiences (prayers, visions), and practices (rituals, approved social behaviors such as fellowship and good deeds). Further, our approach also considers reasons for the metaphysical validity of religion as well as reasons for the truth of specific religious beliefs on their own merits.

Historical Approaches

Most of the historical theories of religion – Freudian, Durkheimian, and even Marxist – share with contemporary theories the underlying naturalist assumption that religion is purely a natural phenomenon and completely explicable in scientific terms. Philosopher and psychologist William James, who invented the field of psychology of religion when he published *The Varieties of Religious Experience: A Study of Human Nature* in 1902, took an evidence-based approach to the academic study of religion. James focused his research on the pragmatic value of religion and particularly on "immediate personal experiences" rather than religious institutions:

> Religion, therefore, as I now ask you arbitrarily to take it, shall mean for us *the feelings, acts, and experiences of individual men in their solitude, so far as they apprehend themselves to stand in relation to whatever they may consider divine.*[8]

[7] Dennett, *Breaking the Spell*, 9.
[8] W. James, *The Varieties of Religious Experience: A Study in Human Nature* (New York: Longmans, Green, and Co., 1902), 31; emphasis original.

James classified religion into two broad types according to the characteristic experiences of their adherents – "religions of healthy mindedness" and "religions of the sick soul." In Saint Francis and his followers, for example, James saw a general optimism and ability to cope with life's circumstances.[9] In other religious personality types, such as that of Martin Luther, he detected unhealthy morbid tendencies to brood over the evil and negative aspects of human life, including one's own.[10]

James was well aware of Sigmund Freud, his contemporary, who was a dedicated atheist. Freud's psychoanalytic method was having documented clinical successes, which the pragmatist James admired. In his 1913 *Totem and Taboo*, Freud presented his psychoanalytic theory that religion originated to meet human needs and thus can be explained in terms of psychological motivations, including wish fulfillment, sexual repression, absolution of guilt, and the need to feel protected against the forces of nature by a powerful father figure. Extending his harsh view of religion in *The Future of an Illusion*, Freud argued that organized religion arose because human persons can live together in mutually beneficial societies only if civilization erects institutions that limit the satisfaction of destructive libidinal drives, such as incest, cannibalism, and murder. As he observed, the religious injunction to "love one's neighbor as oneself" protects civilization from disintegration because such external limits on antisocial instincts become more internalized and religious satisfactions become substitutes for those repressed instincts.

In Freud's *New Introductory Lectures on Psycho-Analysis*, we find the following comment on religion as one phase in the evolutionary trajectory of humanity:

> [R]eligion is an attempt to get control over the sensory world, in which we are placed, by means of the wish-world, which we have developed inside us as a result of biological and psychological necessities .... If one attempts to assign to religion its place in man's evolution, it seems not so much to be a lasting acquisition, as a parallel to the neurosis which the civilized individual must pass through on his way from childhood to maturity.[11]

---

[9] James, *Varieties of Religious Experience*, 80.
[10] James, *Varieties of Religious Experience*, 128, 246.
[11] S. Freud, *New Introductory Lectures on Psycho-Analysis*, trans. W. J. H. Sprott (New York: W. W. Norton and Co., 1933), 229–230.

Interestingly, Christopher Hitchens, a public intellectual and New Atheist, energetically promoted a view like Freud's – that religion is an "infantile" remnant of the primitive in us, which modern persons should replace with reason and science.[12]

Sociologist Émile Durkheim, a contemporary of Freud, rejected Freud's largely individualistic psychological account of religion, partly because humans should have realized by now that we are "victims of error."[13] According to Durkheim's sociological view, religions are found only at the heart of established societies, for the gods connect with (e.g., threaten, bless) the society (tribe, clan, family) rather than the individual. Religion has its origin in social sentiments that tie the individual to society taken as a whole – creating social sentiments and obligations that undergird morality: "Society dictates to the believer the dogmas he must uphold and the rites he must observe; [this indicates] that the rites and dogmas are society's own handiwork."[14] Religion reinforces for the individual his or her role in the larger group by projecting an understanding of the meaning of life from the tribe's perspective. Thus, "religions wholly, or for the most part, [are] a sociological phenomenon" making metaphysical considerations regarding the divine irrelevant.[15] With Durkheim, as with James, we detect the hint of an adaptation hypothesis regarding religion that gained currency later in the twentieth century.

One common theme between nineteenth-century scientific accounts of religion and twentieth-century biological accounts is the belief that a causal account is sufficient. Another commonality is the mixed opinion about whether religion has any positive role to play in modern society. In the early twentieth century, Julian Huxley, biologist and grandson of Thomas Henry Huxley, advocated the putatively scientific view that religion is "a way of life ... which follows necessarily from a man's holding certain things in reverence, from his feeling and believing them to be sacred."[16] Huxley adopted a general evolutionary story that was typical amongst scientific thinkers: religion began in connection to the sacred but has changed over

---

[12] C. Hitchens, *God Is Not Great: How Religion Poisons Everything* (New York: Twelve, 2009), 107–108.

[13] E. Durkheim, "Concerning the Definition of Religious Phenomena," *Durkheim on Religion* (London: Routledge and Kegan Paul, 1975), 94.

[14] Durkheim, "Religious Phenomena," 93.   [15] Durkheim, "Religious Phenomena," 21.

[16] J. Huxley, *Religion without Revelation* (New York: Mentor, 1957), 20.

time, leaving behind emotions of fear to focus more on the rational. Magic, an early stage of religion in which verbal symbols and rituals were used to control unseen forces, gave rise to personification of the forces of nature, attributing human characteristics to supernatural beings who could control these forces for our benefit. Moral standards were later supported by connecting them to the deities. Gradually these deities are unified into fewer gods, resulting eventually in monotheism. For Huxley, religion does have an ongoing role to play in society, as new systems of religious beliefs, made more rational by purging unscientific ideas, can supply helpful guidance and encouragement to assist human flourishing in the further evolution of society and culture.

Philosopher of religion Peter Byrne challenges such revisionary stories and identifies their generally ontological and epistemological antirealist orientation – that religion is not about any transcendent reality or a way of knowing truths about the transcendent.[17] Nonrealist accounts of religion take the view that religious beliefs as well as religion itself are not about what religious believers take them to be about; instead, they arise from subjective psychological experiences or public social experiences. In this regard, evolutionary nonrealist accounts of religion resemble evolutionary nonrealist accounts of morality, giving rise to similar questions about whether such accounts create an epistemic undercutting defeater for religious belief or anchor ontological religious error theory.

Religion as Evolutionary Adaptation

Although scientific accounts of religion may be broadly categorized in evolutionary terms such as adaptationist, nonadaptationist, or maladaptationist, these positions can be connected to different theoretical frameworks that are not necessarily mutually exclusive. The traditional Darwinian framework understands religion as causally driven by biology, that is, selected for its adaptive value. Cognitive accounts of religion see it as produced by cognitive dispositions that have evolved as a result of our basic cognitive capacities, which have adaptive value. Coevolutionary approaches generally accept fundamental Darwinian processes as well as the cognitive mechanisms at work in religion but see culture as the primary evolutionary factor

[17] P. Byrne, *God and Realism* (Burlington: Ashgate, 2003), 11–12.

throughout the history of religion. In this section, we discuss some prominent Darwinian accounts of religion, leaving discussion of the other two approaches for the following sections.

Although Darwinian approaches treat religion as a characteristic of our biology, resulting from natural selection, this does not entail that religion is automatically adaptive, since it would still be thoroughly Darwinian to hold that religion was once adaptive but has since lost its adaptive advantage or even that it always was a nonadaptive by-product of the evolutionary process. What makes a theory of religion Darwinian is that it is premised on natural selection. The major Darwinian accounts these days are adaptationist, asserting that religion is a biologically endowed characteristic or set of characteristics that evolved due to reproductive benefit. Adaptationist theories incorporate a range of proposed advantages religion provides, including reducing the stress induced by the existential fear of death and assistance in attracting mates in sexual selection.[18]

However, the most prominent proposal comes from the eminent evolutionary biologist E. O. Wilson – "the father of sociobiology" – who argues that religion is selection-produced in humans because, much as Durkheim thought, it enhances cohesive group identity:

> The highest forms of religious practice, when examined more closely, can be seen to confer biological advantage. Above all, they congeal identity. In the midst of the chaotic and potentially disorienting experiences each person undergoes daily, religion classifies him, provides him with unquestioned membership in a group claiming great powers, and by this means gives him a driving purpose in life compatible with his self interest.[19]

Regarding cultural evolution, Wilson believes that religious practices that promote reproductive fitness help perpetuate the underlying controlling genes and that religious practices that do not promote fitness (say, practices adopted by the Shakers) lead to the decline of the underlying genes.[20] Yet he holds that biology is still the main cause of religion.

---

[18] Schloss, "Science Unfettered or Naturalism Run Wild?," 20. Examples of the adaptationist approach include R. D. Alexander, *Darwinism and Human Affairs* (Seattle, WA: University of Washington Press, 1979); D. S. Wilson, *Darwin's Cathedral* (Chicago: University of Chicago Press, 2002).

[19] E. O. Wilson, *On Human Nature* (Cambridge, MA: Harvard University Press, 1978), 188.

[20] Wilson, *On Human Nature*, 178.

Given the Darwinian view that religion is produced by natural selection, there is no consensus on whether selection works essentially for the benefit of the group or of the individual. Wilson tends strongly to favor group adaptation, whereas other thinkers in this area favor individual benefit. The idea that religion is a cooperative adaptation at the group level is more established and has generated numerous theoretical and empirical studies, a few of which are enlightening to treat briefly. Two key themes factor into many of the studies: that religion serves to control defections from a cooperative group and that religion functions to coordinate cooperative goals and strategies.[21] Clearly, all cooperative systems depend on capacities for addressing "cheater" control and effecting interactive coordination, whether the cooperative systems are at the level of genomes, multicellular organisms, or social groups.

Costly signaling, examined in Chapter 7, is an excellent example of a behavior that can be a group-level or individual-level adaptation. On the group level, sacrifice by an individual religious believer for the sake of the group is likely to be compensated for by benefits accruing to its group relative to competing groups. As evolutionary psychologist Ara Norenzayan describes it, religion essentially involves "passionate, ritualized communal displays of costly commitments to counterintuitive worlds governed by supernatural agents."[22] Belief in the supernatural thereby has evolutionary adaptive payoff. Evolutionary biologist and anthropologist David Sloan Wilson argues that we must not underestimate the "motivational physiology" of rule enforcement (obey parents, obey priests, obey magistrates) that religion employs to reinforce strong group cohesion.[23] On the individual level, a member of the religious group who engages in costly signaling is more likely to accrue reproductive benefit, which we discuss shortly.

In his groundbreaking work *Darwinism and Human Affairs*, Richard Alexander, a prominent evolutionary biologist, points out that humans have enormously intricate forms of sociality, of a size and complexity that are

---

[21] For example, see D. Johnson and J. Bering, "Hand of God, Mind of Man: Punishment and Cognition in the Evolution of Cooperation," *Evolutionary Psychology*, 4 (2006), 219–233; J. Bulbulia, "Religious Costs as Adaptations that Signal Altruistic Intention," *Evolution and Cognition*, 10 (2004), 19–38.

[22] A. Norenzayan, "Why We Believe: Religion as a Human Universal," in H. Høgh-Olsenen (ed.), *Human Morality and Sociality: Evolutionary and Comparative Perspectives* (London: Palgrave Macmillan, 2010), 60.

[23] Wilson, *Darwin's Cathedral*, 105.

almost nonexistent in the animal world. Cooperation gives advantage in many ways – in gaining food or other important resources and defending against predators, including especially other human groups. Alexander maintains that religion, like morality, was key to creating and sustaining larger, more connected social groups: "[T]he group aspect of religion is a superb unity-producing and maintaining phenomenon."[24] Norenzayan likewise explains that beliefs in supernatural agents allow people to live together in large, cooperative societies far beyond the bounds of small bands of genetically related individuals.[25] Alexander further notes that advanced societies have a morally idealized god that is the negation of the promotion of self-interest or even tribal interest and is the hope of the common person:

> The notion of a god of all people, who was impartial, was the notion of a god who could not condone inequalities such as polygyny, slavery, and caste systems .... This kind of god would have been from the start the hope of ordinary people – the peasants and the downtrodden and the minorities.[26]

Interestingly, Thomas Henry Huxley once commented that a universal idealized god is actually combating natural selection in a certain sense.

Some evolutionary thinkers actually see Jesus's teachings about indiscriminate altruism and self-sacrificial behavior as "reversing evolution," so to speak. Jesus preached love for enemies, sacrificial giving, a worldwide mission, and the moral and spiritual value of caring for the material needs of others. He taught the importance of women, children, the stranger, and the oppressed. Indeed, the universalized "love for neighbor" dissolves all social distinctions, as Saint Paul declared: "In Christ there is neither Jew nor Greek, male nor female, slave nor free."[27] Such values are not rooted in the struggle for survival and natural selection.

Darwinian theories that religion provides something of value and worth to the individual are less well known than group-centered theories, but they deserve mention. Although group-focused and individual-focused approaches to religion need not be mutually exclusive, since selection could operate at both levels, some positions exhibit a degree of exclusiveness. For example, Wilson's group perspective rejects psychological explanations, while physical anthropologist Vernon Reynolds and religion scholar Ralph

---

[24] A. Richard, *The Biology of Moral Systems* (London: Routledge, 2017), 203.
[25] Norenzayan, "Why We Believe," 58–71.   [26] Richard, *The Biology of Moral Systems*, 202.
[27] Galatians 3:28.

Tanner argue that religion benefits only the individual. In evolutionary terms, Tanner and Reynolds focus on how membership in a religious group affects the individual's chance of survival and reproductive success. Employing the r–/r+ distinction in evolutionary biology regarding rates of reproductive activity, they classify organisms with high birth rates as following an r+ strategy and organisms with lower rates as following an r– strategy.[28] Pursuing an r+ strategy – having lots of offspring and providing relatively little parental care (such as in the case of rabbits) – tends to work best in unstable conditions. Organisms pursuing an r– strategy – having fewer offspring and providing more parental care (such as in the case of whales) – tends to work best in stable conditions. Humans, depending on conditions, can exhibit either strategy. As an example of applying this thinking to religion, Tanner and Reynolds find that it is one, but not the only, causally relevant variable accounting for higher birth rates in various predominantly Catholic countries that are less stable and for lower birth rates in many Protestant countries where living conditions are more stable.

An individual-centered approach is also promoted by Dominic Johnson, an evolutionary biologist, and Jesse Bering, an evolutionary psychologist, who argue that natural selection may have favored a widespread human belief in supernatural reward and punishment among our evolutionary ancestors.[29] The major theistic religions contain teachings regarding eternal reward for doing good and eternal punishment for doing evil. In the Muslim scriptures, for example, we find hope of reward: "For them who have done good is the best [reward] and extra. No darkness will cover their faces, nor humiliation. Those are companions of Paradise; they will abide therein eternally."[30] We also find fear of punishment: "Indeed, those who devour the property of orphans unjustly are only consuming into their bellies fire. And they will be burned in a Blaze."[31]

Other psychological approaches emphasize the temporal (not eternal) benefits of religion to the individual, such as a greater sense of well-being or comfort in times of difficulty. Neuroscientist Patrick McNamara argues in *The Neuroscience of Religious Experience* that deep religious propensities have been wired into our cognitive circuitry over evolutionary time, such that

---

[28] V. Reynolds and R. Tanner, *The Biology of Religion* (New York: Longman, 1985), 15ff.
[29] Johnson and Bering, "Hand of God, Mind of Man," 219–233.   [30] Qur'an 10:26.
[31] Qur'an 4:10.

religion has been a defining mark of what it means to be human.[32] He even argues that the normal function of religion is to support cognitive processes that lead the "self" to aspire to become an "ideal self," as defined by that religion.[33]

Among biological accounts of religion as an evolutionary adaptation, religion receives a reasonably positive, or at least neutral, treatment because of its envisioned adaptive roles. Nevertheless, as a causal explanation of religion, a biological account can be construed as totalistic, completely explaining our religious tendencies and undercutting any rationale for holding religious beliefs. Of course, this posture toward religion is a philosophical interpretation of science rather than science per se, based on a move from the ostensibly neutral approach of methodological naturalism to metaphysical naturalism and strict empiricism.

Does explaining the *capacity* for forming religious beliefs without reference to the *content* of the beliefs provide an epistemic defeater for religious beliefs about the supernatural? Is the plausibility of explanation that cites known biological factors a reason to conclude that there are no other factors? Obviously, naturalists working on these issues think so. However, some Christian theists point out that Christian doctrine affirms the physical conditions of human life, which entails that higher capacities, such as rational thought, moral awareness, and a sense of the divine, will be mediated through biological realities. Psychologist Justin Barrett, a Christian theist, argues that cognitive science reveals "some particular features of human minds that make belief in superhuman agents natural" – that, from childhood, we humans have a general innate tendency to believe in god. This "cognitive architecture," as he calls it, can readily be seen to be the result of brain evolution. Anthropologist and atheist Scott Atran makes the same point in his book *In Gods We Trust*.[34]

Barrett first observes that the general innate tendency to believe that other persons have minds seems to have been built into our cognitive architecture. He asks rhetorically whether a complete scientific explanation for why humans nearly universally believe that other people have minds

[32] P. McNamara, *The Neuroscience of Religious Experience* (Cambridge: Cambridge University Press, 2009), ix.

[33] McNamara, *Neuroscience of Religious Experience*, xi; see also 44.

[34] S. Atran, *In Gods We Trust: The Evolutionary Landscape of Religion* (New York: Oxford University Press, 2002), 57.

would suddenly count against that belief – that is, against whether humans *should* believe in other minds. Thus, says Barrett,

> Belief in other minds and belief in gods are both highly intuitive consequences of cognitive architecture operating on ordinary inputs. Both are non-empirical, widespread beliefs. Neither is directly weakened by increasing scientific knowledge about how these beliefs come about any more than knowledge of the visual system makes us suspicious that the visual world is not really out there.[35]

Barrett concludes that to point out the scientific fact that the structure of the evolved human mind encourages certain beliefs is hardly to raise a sovereign objection to the truth and rationality of those basic beliefs. Many Christian thinkers argue, for example, that Christianity predicts that God, who seeks relationship with us, would create a world in which a capacity for awareness of him would evolve. These thinkers then state that a world with creatures capable of seeking God is exactly what we have. In Romans, Saint Paul writes, "that which is known about God is evident within them; for God has made it evident to them."[36] To the extent that we have scientific facts about the evolution of religious capacities, we have one more case where theists and nontheists offer conflicting interpretations of those facts.

The epistemological work of Christian philosopher Alvin Plantinga resonates with Barrett's psychological work in holding that belief in God is an almost inevitable consequence of the kinds of minds we have. Plantinga explains that Christian theism entails that God willed rational creatures to exist and develop a panoply of belief-forming powers aimed at truth. In appropriate circumstances, these powers activate to form rationally warranted beliefs immediately and directly without discursive argument. Our beliefs in other minds, the external world, and the past just begin the list of immediate beliefs that are so fundamental that they cannot be argued. Among the various human belief-forming powers, such as perception and memory, is also an innate disposition to believe in God – which Plantinga designates using John Calvin's term *sensus divinitatis*, a sense of the divine. Calvin writes, "There is within the human mind, and indeed by natural instinct, an awareness of divinity . . .. [T]his conviction, namely that there is

[35] J. Barrett, "Cognitive Science, Religion, and Theology," in Schloss and Murray (eds.), *Believing Primate*, 96.
[36] Romans 1:19.

some God, is naturally inborn in all, and is fixed deep within, as it were in the very marrow."[37] Plantinga concludes, then, that the epistemological warrant for beliefs, which occur naturally out of the human constitution, is best supported by supernaturalist metaphysics.[38]

Religion as Cognitive Incidental

In the cognitive science of religion (CSR), religion is viewed as a by-product of the evolutionary process – a "spandrel" – in which case it is nonadaptive.[39] CSR – which combines cognitive, developmental, and evolutionary psychology with anthropology and history of religion – got traction in the 1990s when Pascal Boyer published *Tradition as Truth and Communication* and *The Naturalness of Religious Ideas*. His fundamental idea was that religion involves native cognitive dispositions that have evolved along with our cognitive capacities. In his 2001 *Religion Explained*, Boyer writes, "The building of religious concepts requires mental systems and capacities that are there anyway, religious concepts or not."[40] Our cognitive capacities have adaptive value, whereas religion as an associated disposition does not have adaptive value, although it may have been adaptive in the ancestral environment.

Several specific proposals have been offered on how religion is an evolutionary incidental – from religious ritual as harm avoidance behavior to religious emotion as expression of longing for attachment.[41] However, the salient theory asserts that humans have the innate tendency to attribute intentions and agency to a wide variety of things in their environment. Anthropologist Stewart Guthrie argues in *Faces in the Clouds* that humans have an almost universal propensity for intentional attribution. Guthrie asserts that this bias underlies animistic religion and is expressed in more developed religions, perhaps even explaining why people are susceptible to the argument from design. Yet, as Guthrie put it, "the clothes have no

---

[37] John Calvin, *Institutes of the Christian Religion*, 1.3.1–3.

[38] A. Plantinga, *Warrant and Proper Function* (New York: Oxford University Press, 1993), 237.

[39] The term was brought into biology by S. J. Gould and R. Lewontin, "The Spandrels of San Marco and the Panglossian Paradigm," *Proceedings of the Royal Society, Series B*, 205, no. 1161 (1979), 581–598.

[40] P. Boyer, *Religion Explained: The Evolutionary Origins of Religious Thought* (New York: Basic Books, 2001), 331.

[41] R. N. McCauley and E. T. Lawson, *Bringing Ritual to Mind* (New York: Cambridge University Press, 2002); K. J. Eames, *Cognitive Psychology of Religion* (Long Grove, IL: Waveland Press, 2016), 117–133; S. Guthrie, *Faces in the Clouds* (Oxford University Press, 1993).

emperor."[42] Evolutionary psychologist Paul Bloom goes further in suggesting that we are all natural-born Creationists.[43] Since Creationism is particularly popular in the United States, a number of studies have been conducted to demonstrate the innate propensity in children to attribute creative agency to inanimate things (such as a neat pile of blocks) while the adults studied predictably differed according to their educational background and cultural environment.[44]

Daniel Dennett, philosopher of cognitive science and New Atheist, makes the case that religion arose when people embraced an *intentional stance*, an animal response that attributes agency to a wide variety of things in the environment, animate or inanimate, that puzzle or frighten. Dennett asserts that the capacity for the intentional stance is rooted in innate mechanisms or capacities for agency detection that evolved in higher animals, but with a bias to make false positives in attributing agency to nonagents. Cognitive scientists call this capacity a "hypersensitive agency detection device" (HADD).[45] From an evolutionary perspective, Dennett calls this the "Good Trick" – an impressive feature of the human mind – which has the ability to preserve our lives by treating things as agents with beliefs and intentions.[46] Religion, according to Dennett, makes use of this capacity.

A significant amount of empirical work undergirds this theory for animals while also showing humanity's nearly universal tendency for anthropomorphically projecting supernatural agents, like gods and demons, onto the natural world. Primitive people might say, for instance, that a severe thunderstorm results from angering the gods. Logically, these studies do not show that religion has no adaptive functions, but they do establish an important cognitive tendency that has deep evolutionary roots.[47]

Dennett observes that religion as a human phenomenon is a hugely costly endeavor, requiring the expenditure of energy and resources, leading him to state that something must pay for it. He quotes the Latin, *cui bono?* – "Who

---

[42] Guthrie, *Faces in the Clouds*, 5.

[43] P. Bloom, "Religious Belief as an Evolutionary Accident," in Schloss and Murray (eds.), *Believing Primate*, 121.

[44] F. Heider and M. Simmel, "An Experimental Study of Apparent Behavior," *American Journal of Psychology*, 57 (1944), 243–259; G. Csibra et al., "One-Year-Old Infants Use Teleological Representations of Actions Productively," *Cognitive Science*, 27 (2003), 111–133; Guthrie, *Faces in the Clouds*.

[45] J. Barrett, "Exploring the Natural Foundations of Religion," *Trends in Cognitive Science*, 4 (2000), 29–34.

[46] Dennett, *Breaking the Spell*, 67, 108–114.   [47] Guthrie, *Faces in the Clouds*.

benefits from this?" Evolutionary biology teaches that any persistent phenomenon in the living world that apparently exceeds its function requires explanation because evolution is so efficient in eliminating pointless accidents. Dennett answers the question by indicating that religion survived because it developed ideas that could be culturally inherited in connection with other genetically based cultural ideas: "Believe and obey your parents" protects children against danger. Selection favored the emergence of a trusted "information superhighway" between parents and children through which cultural ideas and religious stories could pass. Religion, then, uses the idea of a trusted supernatural parent – especially a "Father."

Dennett adopts Richard Dawkins's view that the survival of religion in human culture is bound up with "memes" – units of cultural transmission, rooted in the brain or cultural artifacts, much as genes are units of biological inheritance. Thus, memes (Greek: *mimesus*, imitation) or memeplexes replicate and transmit these cultural ideas. Rituals such as prayer, singing, dancing, and reciting where people act in unison are particularly powerful ways to transmit memes. Eventually, gene stewards – shamans, priests, and ministers – appeared in the history of religion to shepherd the memes because they personally had something to gain from their preservation.

Dennett's evolutionary explanation of religion amalgamates ideas from biology, cognitive science, and coevolutionary theories of the role of culture interacting with biology. Like anthropologist James George Frazer and Freud, he recognizes that religion is socially significant and can foster positive traits, such as kindness toward others or the resolve to abstain from drugs, but he also speculates that perhaps there might be atheist groups that do the same positive things. Besides, as Dennett points out, religion can foster negative traits, such as bigotry and enforcement of ignorance, which would make it on balance a negative in human life.

Religion as Maladaptive Memetic Pathogen

No evolutionary account of religion is more famous than Richard Dawkins's theory that religion is maladaptive, a harmful human phenomenon that is the product of biology and culture. While there is considerable debate among evolutionary theorists and philosophers of biology over what it means to be adaptive, our discussion connects it to fitness. Thus, the basic question is whether religion generates behaviors that enhance human fitness. However, we recognize prior questions that we cannot pursue at length, such as, What is human fitness? Should human fitness be linked to a broad conception of human flourishing? Jeffrey Schloss asks how we could ever measure fitness or flourishing. Should we count the total number of religious adherents, or determine religion's net influence on individual lives on some hedonic or eudaemonic index, or weigh the all-things-considered impact of religion on the trajectory of human history, or what?[48] Further, if adaptivity in religion is not strictly binary, then we must discriminate which religions and which religious practices are and are not adaptive, by whatever measure.

Dawkins laid out his coevolutionary account in 1976 in *The Selfish Gene*, which acknowledged that culture plays a key role in archiving and transmitting information that informs the behavior of individuals and groups. His neologism – "meme" – refers to the basic unit of cultural storage, replication, and transmission.[49] A meme conveys an idea or behavior or theme that spreads from person to person within a culture, often with the assistance of writing, speech, song, and rituals. "Examples of memes," he suggests, "are tunes, ideas, catch-phrases, clothing fashions, designs for pots or building arches."[50] Actually, as Dawkins further clarifies, these are meme cultural products, meme phenotypes – not the instructions but the results – just as phenotypical expressions of genes, like hair color or wing length, are biological gene products rather than the genetic instructions. Ostensibly residing in the brain, memes as units of information or instruction self-replicate, mutate, and respond to selective pressures.

Dawkins's selfish-meme theory parallels his selfish-gene theory. As we saw in previous chapters, Dawkins employs selfish-gene theory to explain a wide variety of phenomena, from genetic propagation to morality, as ultimately existing to preserve the selfish gene. Just as a gene is an automatic replicator, so is the meme:

> Once the genes have provided their survival machines with brains that are capable of rapid imitation, the memes will automatically take over. We do not even have to posit a genetic advantage in imitation, though that would

[48] Schloss, "Science Unfettered or Naturalism Run Wild?" 16.
[49] R. Dawkins, *The Selfish Gene* (New York: Oxford University Press, 1976), 249.
[50] Dawkins, *The Selfish Gene*, 249.

certainly help. All that is necessary is that the brain should be capable of imitation: memes will then evolve that exploit the capability to the full.[51]

For Dawkins, religions are "the prime examples of memes,"[52] exploiting throughout evolutionary history the meme's striking capability to replicate.

Dawkins's 1991 essay "Viruses of the Mind" describes religious beliefs as "mind-parasites" and believers as "faith sufferers" or "patients." After all, physical sickness gets passed from one individual or population to another, why not mind viruses? Dawkins proceeds to critique two harmful traits of religion interpreted as viruses of the mind. First, Dawkins insists, memes, particularly religious memes, can bypass normal rational processes:

> The patient typically finds himself impelled by some deep, inner conviction that something is true, or right, or virtuous: a conviction that doesn't seem to owe anything to evidence or reason, but which, nevertheless, he feels as totally compelling and convincing. We doctors refer to such a belief as "faith."[53]

Although logic and evidence are virtues in normal life, Dawkins thinks that religion makes virtues of lack of logic and absence of evidence.

Second, Dawkins charges that religious faith is harmful and destructive, a source of bigotry and violence, "one of the world's great evils, comparable to the smallpox virus but harder to eradicate."[54] Neuroscientist Sam Harris, another New Atheist, echoed similar sentiments about religious violence in *The End of Faith: Religion, Terror, and the Future of Reason*, which he published shortly after the September 11, 2001, terrorist attacks on the United States. Harris declared that "religious faith perpetuates man's inhumanity to man."[55] Increasingly, the topics of religious intolerance and violence are the subjects of scientific conferences, books, and research.[56]

---

[51] Dawkins, *The Selfish Gene*, 259, 328.

[52] R. Dawkins, *A Devil's Chaplain: Reflections on Hope, Lies, Science, and Love* (Boston: Houghton Mifflin, 2003), 117.

[53] R. Dawkins, "Viruses of the Mind," in B. Dahlbom (ed.), *Dennett and His Critics* (Oxford: Blackwell, 1993), 20.

[54] R. Dawkins in a speech delivered to the American Humanist Association, accepting the award for 1996 Humanist of the Year. Published as "Science Verses Religion," in L. P. Pojman and M. Rea (eds.), *Philosophy of Religion: An Anthology*, 5th ed. (Belmont, CA: Thomson Wadsworth, 2008), 426.

[55] S. Harris, *The End of Faith: Religion, Terror, and the Future of Reason* (New York: W. W. Norton, 2005), 14.

[56] S. Clarke, R. Powell, and J. Savulescu (eds.), *Religion, Intolerance, and Conflict: A Scientific and Conceptual Investigation* (Oxford: Oxford University Press, 2013).

Both of Dawkins's criticisms discussed earlier deserve measured reflection. In addition to his first point that religion is perpetuated by memes rather than by rational thought, he is especially known for deconstructing William Paley's analogical design. In *The Blind Watchmaker*, he seeks to demonstrate the rational insufficiency of religion by showing that natural selection can produce organized, complex biological structures that seem designed without the work of supernatural intelligence.[57] Dawkins, Dennett, and others who share this opinion often make sweeping indictments of the theistic arguments but do so without a high degree of technical philosophical engagement. Most intellectually sophisticated theists recognize the weakness of Paley's argument and yet give credence to better forms of teleological argumentation – such as the fine-tuning and anthropic arguments, which are scientifically informed.[58] The Society of Christian Philosophers, founded in 1978, is often cited as a case study of the intellectual vibrancy of theistic and Christian belief, because the society engages believers and nonbelievers alike in rational dialogue regarding the rational credentials of Christian faith.[59] An example of philosophical engagement between both sides is found in *Science, Evolution, and Religion*, a debate by Christian theist Michael Peterson and Darwinian naturalist Michael Ruse, in which both stipulate the truth of evolutionary theory but differ in their worldview interpretations of it.[60]

As a causal explanation of religious cognition and behavior, meme theory can motivate both ontological and epistemological antirealism about religion – that it is not about anything real and not a way of knowing truth. Many scientists think that the idea of memes as mechanisms of cultural transmission is inadequately defined, either operationally or empirically, and thus that it must await quantified experimental support.[61] Additionally, some religious believers have asked why atheism is not itself a meme that replicates through culture; but to call a belief a meme is supposedly to undercut its rationality.

[57] R. Dawkins, *The Blind Watchmaker* (New York: W. W. Norton, 1986), 6.
[58] M. L. Peterson, et al., *Reason and Religious Belief: An Introduction to the Philosophy of Religion*, 5th edn. (New York: Oxford University Press, 2013), 93–101.
[59] See the multi-decade contributions in *Faith and Philosophy: Journal of the Society of Christian Philosophers*, which is open access at www.faithandphilosophy.com. See also www .societyofchristianphilosophers.com.
[60] M. L. Peterson and M. Ruse, *Science, Evolution, and Religion: A Debate about Atheism and Theism* (New York: Oxford University Press, 2016).
[61] For example, see J. Coyne, "The Self-centered Meme," *Nature*, 398 (1999), 767–768.

For instance, scientist and Christian theologian Alister McGrath puts it this way: "If all ideas are memes, or the effects of memes, Dawkins is left in the decidedly uncomfortable position of having to accept that his own ideas must also be recognized as the effects of memes."[62] Since it is close to incoherent to think that theism and atheism are equally memes, and thus equally valid or invalid, it appears that more direct intellectual engagement between religion and its critics is required.

In pursuing Dawkins's second point – that religion is harmful – one might ask whether all religions are toxic or whether some religions are beneficial. Religions vary widely throughout history – from primitive animist religions to religions with well-developed moral practices and theological systems. Should they all be painted with the same brush? For example, there are a great many religious charitable organizations with humanitarian outreach – feeding the needy, offering agricultural and medical assistance in under-developed countries, founding and supporting major colleges and universities, and the like. Some would argue that, in many religions, negative behaviors among their adherents are simply the failure to live up to the religion's high ideals rather than a genuine expression of them. Interestingly, Susan Blackmore, an expert in memetics, who once agreed with Dawkins that religion is a harmful cultural virus, later came to believe that religion could be beneficial as well as harmful.[63]

Even as a naturalist reductionist offering a coevolutionary interpretation of religion, Dawkins has acknowledged that humans are in some sense unique.[64] As Dennett remarks, "We have creeds, and the ability to transcend our genetic imperatives. This in fact makes us different."[65] The theist might well argue that a naturalist foundation makes it difficult to support any credible claim about possible human uniqueness but that a religious and theological foundation can support such a claim. For instance, the orthodox Christian claim, asserted with scientific awareness, is that God willed an evolutionary world that would eventually bring forth

[62] A. E. McGrath, *Dawkins' God: From the Selfish Gene to the God Delusion*, 2nd edn. (Oxford: Wiley Blackwell, 2015), 126.

[63] S. Blackmore, "Why I No Longer Believe Religion Is a Virus of the Mind," *The Guardian*, September 16, 2010, www.theguardian.com/commentisfree/belief/2010/sep/16/why-no-longer-believe-religion-virus-mind.

[64] Dawkins, *The Selfish Gene*, 245.   [65] Dennett, *Breaking the Spell*, 4.

rational–moral–social creatures capable of transcending their biology and developing an awareness of him.[66]

Religious Accounts of Biology

How should we think about the various evolutionary explanations of religion? Although they fall under the rubric of science, they vary widely in their degree of empirical grounding, some citing experiments performed and others reflecting tacit worldview assumptions. Currently, in the scientific study of religion, disagreements among theories are obvious, such that they cannot all be true. Further, available scientific explanations currently emanate from at least three theoretical frameworks, one holding that biology is fundamental, another that it is a by-product, and still another that it has been superseded by culture. It is not just theories of religion that are diverse; religions themselves are complex and multifaceted, involving beliefs, rituals, experiences, emotions, life practices, and social interactions that are hard to bring under a single comprehensive theory. Should science focus largely on seemingly universal religious beliefs – or should it engage the particularities of individual religions? Greater clarity is yet to come as CSR seeks to settle these and other questions.

What is clear is that the search for causal explanations of religion has found new conceptual resources within several theoretical evolutionary frameworks. While officially claiming scientific neutrality about religious beliefs, many thinkers produce explanations of religion that exclude any explanation for the truth and plausibility of religious beliefs. Philosophers Alex Rosenberg and Tamler Somers straightforwardly state that "if our best theory of why people believe P does not require that P is true, then there are no grounds to believe P is true" – a principle that, if correct, justifies using evolutionary explanation to dismiss any rationale for religious belief.[67] Take theistic belief, for example. Many in the biological study of religion claim that theism is unlikely to be true, given some preferred biological account.

---

[66] Peterson takes this position in Peterson and Ruse, *Science, Evolution, and Religion*, 142. See also M. Murray, "Scientific Explanations of Religion and the Justification of Religious Belief," in Murray and Schloss (eds.), *Believing Primate*, 168–178.

[67] T. Sommers and A. Rosenberg, "Darwin's Nihilistic Idea: Evolution and the Meaninglessness of Life," *Biology & Philosophy*, 18 (2003), 667.

Or, where T is the proposition that theism is true, and B is a particular biological account, the judgment is that P (T/B) is low.

Since many who offer biological accounts of religion hold naturalist worldview commitments, theists raise the possibility that those accounts are influenced by their naturalism. Hence, theists can raise the possibility that the negative assessment of theism is based on N – that is, naturalism – conjoined with B, which is the preferred biological account. If this is the case, then a more accurate symbolic portrayal of the assessment would be as follows:

$$P(T/B \; \& \; N) \text{ is extremely low.}$$

For many theists, the implicit appeal to naturalist background assumptions makes the low probability assessment for theism in the name of science problematic.

Let us briefly switch our emphasis from biological accounts of religion to religious accounts of biology. More specifically, we might ask which world-view – theism or naturalism – makes better sense of existence and the findings of biology itself. In what kind of universe is the science of biology more likely to arise? The theist can point out that theism entails that the cosmos exists by the will of a supremely powerful creator and has order because of his supreme wisdom. The theist can press further, indicating that naturalism implies that the universe exists purely by chance and has the lawlike order it has purely by chance. Since the necessary conditions for science, including biological science, include a rationally ordered world and rational inquirers, the theist can claim that science, including biological science, is much more likely to arise in a theistic universe than in a naturalistic universe. Or, where P is antecedent probability, S is science, T is theism, and N is naturalism, we may formalize the claim as follows:

$$P(S/T) > P(S/N)$$

Likewise, the theist can claim that, given theism, the science of biology is for these same reasons more likely to arise.

Now, disagreements among theoretical frameworks, concrete proposals, and even background philosophical assumptions are not necessarily a bad thing in a relatively new field such as CSR. No doubt, more creative and insightful theoretical and empirical work will be forthcoming. On the philosophical level, all involved must be aware of how worldview perspectives can influence claims made for the scientific study of religion. To illustrate this point: consider that Wilson states that "theology is not likely to survive" the rational power of scientific materialism, while Plantinga maintains that theological truths provide the only adequate description of the kind of reality in which life, rational thought, and science make sense.[68] Opposing world-views will always be at odds over religion, which means that the new territory to be charted by scientific advances in the study of religion will in turn be subject to philosophical debate.

[68] Wilson, *On Human Nature*, 192. Plantinga, *Warrant and Proper Function*, 237.

10 Humanity, Religion, and the Environment

Over the past several decades, the connections between ecology and all major religious perspectives, including the Abrahamic traditions, have been extensively explored. Due to such explosive growth in theological scholarship, official institutional commitments, and public action by the religious community, the environment has become an important topic in science–religion scholarship.

In the present chapter, we briefly explore the history of the relation between religion and the environment. We then move to a fuller discussion of specifically monotheistic religion in this regard, including claims that it is a major cause of environmental degradation as well as counterclaims that it can be an important part of the solution. We also discuss some other major perspectives that seek positive engagement with environmental concerns, from movements seeking spiritual identification with nature to the atheist groups that support protecting the environment on other grounds. We close the chapter by looking at the complex relations between religion, the environment, and politics.

Religion, Nature, and Ecology

Discussion of the relation of religion to the environment must logically be prefaced by inquiry into religious understanding of nature itself. Of course, understandings of nature have differed among religions throughout history and around the globe, making a brief survey important for perspective on our subject. After briefly considering selected religions of the distant past, we then turn to how contemporary religions think about nature and the environment.

Humanity's earliest identifiable religion was *animism*, which was first defined and studied in the nineteenth century by E. B. Tylor, the founder of anthropology.[1] Animism as a tribal religion, originating when humans were hunter–gatherers, holds the fundamental belief that nature is pervaded by some force or power. From the Latin *anima* – meaning spirit or life – animism is nondualistic in seeing the natural world as alive somehow and in kinship with humanity. Since animism is preliterate and prescientific, it has produced no complex theologies that sharply distinguish nature and humanity; rather, there is simply the treatment of nature as alive. Subsequent studies on the origin of religion made similar generalizations based on two sources – inference from archeological finds and observations from contemporary cultural anthropology.[2] Fascinatingly, many groups in New Paganism explicitly represent the "new animism," particularly by encouraging respect for nature and thus interest in environmental issues.[3] Religious studies expert Harvey Graham notes that renewed academic interest in animism and ecology challenges dominant religious ideologies and practices in the developed world.

Hinduism, which began perhaps as long as 5,000 years ago, was rooted in nature worship that involved a pantheon of gods who were identified with various aspects of nature, such as Indra with storms and Varuna with the sky. As Hindu religion evolved, it was eventually given philosophical expression in the fundamental insight of the *Upanishads* that the individual self is the Great Self: *Atman* is *Brahman*. Classically, the teaching that Brahman is the substrate of everything and pervades everything was connected to the idea that the physical world is a realm of illusion (Sanskrit: *maya*), which must be overcome and dispelled by achieving higher insight. This otherworldly, antimaterialist view of nature may make us wonder whether Hinduism contains the conceptual resources for positive engagement with environmental issues. However, the *Śrīmad Bhāgavatam* teaches that the "[e]ther, air, fire, water, earth, planets, all creatures, directions, trees and plants, rivers and seas … are all organs of God's body."[4] This theme underwrites the Hindu idea that respect and reverence toward all

[1] E. B. Tylor, *Primitive Culture: Researches into the Development of Mythology, Philosophy, Religion, Art, and Custom*, 2 vol. (London: John Murray, 1871), vol. I, 377–453.

[2] N. Smart, *The Religious Experience of Mankind* (New York: Charles Scribner's Sons, 1969), 49–54; G. Harvey, *Animism: Respecting the Living World* (New York: Columbia University Press, 2006), 3–29.

[3] G. Harvey, "Animist Paganism," in M. Pizza and J. R. Lewis (eds.), *Handbook of Contemporary Paganism*, Brill Handbooks on Contemporary Religion (Leiden: Brill, 2009), 401.

[4] *Śrīmad Bhāgavatam* 11.2.41.

living beings is required and will lead to harmonious relationship between humans and nature.

Reverence for living beings anchors the idea of *ahimsa*, or nonviolence, which has important implications for environmental thinking and attitudes. As India underwent seemingly unbridled industrialization in the twentieth century along with the problems that accompanied it, Mahatma Gandhi offered a critique:

> This land of ours was once, we are told, the abode of the Gods. It is not possible to conceive Gods inhabiting a land which is made hideous by the smoke and the din of mill chimneys and factories, and whose road ways are traversed by rushing engines dragging numerous cars crowded with men who know not for the most what they are after.[5]

He also pointed out that the direction for a solution to the problem is a nonviolent economy:

> Strictly speaking, no activity and no industry is possible without a certain amount of violence, no matter how little . . .. What we have to do is to minimize it to the greatest extent possible.[6]

Gandhi broadened the principle of nonviolence to "nonviolent resistance" in his quest for Indian independence, but many believe that the principle has continued to guide environmental advances in India, which include slowing deforestation and encouraging organic farming. In 1990, Indian social activist Baba Amte, following Gandhi's approach, conducted a vigil to prevent the planned destruction of forest and agricultural land by the construction of the Narmada River Valley dam project.[7] Indologist Christopher Chapple reports that in the past several decades India has made encouraging efforts toward environmental awareness, often led by the Gandhi Peace Foundation and the Center for Environment Education.[8]

After Buddhism broke from Hinduism in the sixth century BC, it eventually spread throughout Southeast Asia and became a major world religion. A Buddhist response to the environmental crisis may be developed from

---

[5] M. K. Gandhi, *My Socialism*, R. K. Prabhu (ed.) (Ahmedabad: Navajivan Publishing House, 1959), 32.

[6] Gandhi, *My Socialism*, 32.

[7] C. Chapple, "Contemporary Jaina and Hindu Responses to the Ecological Crisis," in Michael Barnes (ed.), *An Ecology of the Spirit: Religious Reflections and Environmental Consciousness* (Lanham, MD: University Press of America, 1990), 216.

[8] Chapple, "Contemporary Jaina and Hindu Responses to the Ecological Crisis," 209–218.

several key ideas, but the core concept of "desire" provides the starting point. Siddharta Gautama – who became "the Buddha," the Enlightened One – taught in his Four Noble Truths that suffering is caused by the inappropriate desire to cling to the things of this world.[9] In this light, environmental harm is due to wrong desire. The Buddha's teachings also included the principle of interdependence – the idea that everything is interdependent with every-thing else, in a vast causal web. Writing on Buddhist environmental ethics, Rita Gross, an American Buddhist feminist scholar of religions, explains how this idea counteracts Western individualism:

> Simply put, interdependence means that nothing stands alone apart from the matrix of all else . . . . Given interdependence, our very identity as isolated, separate entities is called into serious question and we are invited to forge a more inclusive and extensive identity . . . . Western Buddhists especially seem to find immense relief in their discovery of [the concept of] the "we-self."[10]

The Buddhist critique targets corporate, national, and personal greed and consumption as major spiritual problems, since their impact intervenes in or rearranges ecosystems and thus damages balanced interconnectedness.

Buddhist meditation is particularly aimed at dealing with the fundamental problem of the self's inherent selfishness (Sanskrit: *trishnā*) that is exhibited in attachment, clinging, craving, and fixating in relation to things and agendas. Applied to environmental concern, inordinate desire causes misery (Sanskrit: *duhkha*), which can be manifest in overconsumption and overpopulation, which are mistakenly sought to bring satisfaction. However, the spiritual goal is for the grasping self (the "I want …") to be replaced by enlightened detachment and contentment. On the environment, Buddhism seeks to trans-form the self in relation to nature and to help others experience the same transformation – particularly in regard to noninjury – which means, put positively, "loving kindness" toward all beings in an interconnected world.

As the Dalai Lama has urged, "the inner environment" of our mental awareness can change and heal the outer environment.[11] Various contem-porary Buddhist thinkers are working to develop a Buddhist ecology that

---

[9] Samyutta Nikaya 56.11.

[10] M. R. Gross, "Toward a Buddhist Environmental Ethic," *Journal of the American Academy of Religion*, 65, no. 2 (1997), 337–338.

[11] Dalai Lama, "Universal Responsibility and the Inner Environment," presented at the University of Portland on May 9, 2013. Available online at www.dalailama.com/videos/universal-responsibility-the-inner-environment.

joins the discussion among the religiously committed, socially concerned people of the world. It is important to note that, besides asserting that life is full of suffering due to improper desire that may be eliminated by following Buddhist teaching and practice, Buddhism projects no overall purpose to life; it thereby repudiates all forms of monotheism that assert a purpose for humanity and nature as inconsistent with the Buddhist perspective.

The field of ecology and religion continues to grow at a rapid pace. There are numerous conferences, publications, and organizations themed around the need to understand human cultures, religions, and environments. The *Encyclopedia of Religion and Nature* and the *Oxford Handbook of Religion and Ecology* have become important reference works that bring definition and order to our understanding of this multifaceted phenomenon.[12] The Religion and Ecology group, which was launched in the early 1990s under the auspices of the American Academy of Religion, represents another concerted effort to focus scholarly attention on religion in human/ecosystem interactions and to explore whether dominant religions might be mined for positive environmental principles.

Monotheism and the Environment

Monotheism – the view that one Supreme God created everything else – arose in the ancient Near East within an overall context dominated by polytheistic religions. Polytheism is the belief in many gods who are each identified with a certain area of nature, such as the sun or moon. Ancient Sumeria, Babylonia, and Egypt, for example, were all polytheistic cultures long before Greece and Rome, whose polytheistic religions became subjects of classical literary mythology. It was into this pervasively polytheistic environment that monotheism originated as a sharply contrasting view of the divine. The first chapter of Genesis is appropriately read as a polemic against surrounding religions – for example, Genesis 1:3 may be paraphrased as follows: "[T]he sun and the moon are not gods, as the surrounding polytheistic religions believe, for the one God created both."[13] The patriarch Abraham

[12] B. Taylor (ed.), *Encyclopedia of Religion and Nature*, 2 vols. (New York: Continuum, 2005); R. S. Gottieb (ed.), *Oxford Handbook of Religion and Ecology* (New York: Oxford University Press, 2006).

[13] J. K. Hoffmeier, "Some Thoughts on Genesis 1 and 2 and Egyptian Cosmology," *Journal of the Ancient Near Eastern Society* 15 (1983), 39–49. See also G. F. Hasel, "Polemic Nature of the Genesis Cosmology," *The Evangelical Quarterly*, 46 (1974), 81–102.

had faith that a new people, the Hebrews, would be raised up to know and obey God; Moses received the Ten Commandments from God on Mount Sinai, the first of which affirmed the sole God's rightful position and forbade the worship of false gods. Christianity branched from this tradition in the first century AD, and then Islam arose from it in the seventh century AD.

The Abrahamic faiths – Judaism, Christianity, and Islam – share the belief that everything that is not God is creature, both humanity and nature, and that everything in creation has a purpose in God's plan. In the Judeo-Christian tradition, humans are made in "the image of God"[14] – being endowed with finite rational, moral, and relational capabilities resembling God's infinite attributes. The Genesis account of creation depicts God giving humans oversight of nature: "[Y]ou are to have dominion over the fish of the sea, and over the fowl of the air, and over every living thing that moves upon the earth."[15] Psalm 8 echoes this general understanding in giving praise to the Lord:

> What is humankind that you are mindful of them,
>   human beings that you care for them?
> You have made them a little lower than angels
>   And crowned them with glory and honor.
> You made them rulers over the works of your hands;
>   You put everything under their feet:
> all flocks and herds,
>   and the animals of the wild,
> the birds in the sky,
>   and the fish in the sea,
>   all that swim the paths of the seas.[16]

However, environmental thinkers have criticized the cultural ideas and attitudes toward nature that the biblical account has historically supported.

In 1967, historian Lynn White published "The Historical Roots of Our Ecologic Crisis," a short paper that blamed Christianity for environmental problems because for centuries it has fostered a hierarchical view of reality in which humans are superior to nature. He developed the point that the religious perspective that humans have dominion over nature was taken to imply that nature is purely for human use:

---

[14] Genesis 1:27.   [15] Genesis 1:28.   [16] Psalm 8:4–8.

Especially in its Western form, Christianity is the most anthropocentric religion the world has seen .... Man shares, in great measure, God's transcendence of nature. Christianity, in absolute contrast to ancient paganism and Asia's religions (except, perhaps, Zoroastrianism), not only established a dualism of man and nature but also insisted that it is God's will that man exploit nature for his proper ends .... By destroying pagan animism, Christianity made it possible to exploit nature in a mood of indifference to the feelings of natural objects .... The spirits *in* natural objects, which formerly had protected nature from man, evaporated. Man's effective monopoly on spirit in this world was confirmed, and the old inhibitions to the exploitation of nature crumbled.[17]

White's target was particularly Christian thinking about nature that emerged in the late medieval world and set the stage for the Scientific Revolution in the West, the Industrial Revolution, and the resulting explosion in technology. Not only did certain Christian ideas provide the context for science and technology to originate and develop, they supported a zeitgeist that encouraged exploitation of the environment. Clearly, for White, what it means to "have dominion" was interpreted in a negative, dualistic, chauvinistic way toward nature, leading to devastating ecologic effects. Thus, he charges, "Christianity bears a huge burden of guilt."[18]

White urges the Christian community to reject the "axiom that nature has no reason for existence save to serve man"[19] and to recover an alternative Christian perspective, one demonstrated by St. Francis of Assisi, who taught about the intimate relation between humanity and nature. "Both our present science and our present technology," White argues, "are so tinctured with orthodox Christian arrogance toward nature that no solution for our ecologic crisis can be expected from them alone." He continues, "[S]ince the roots of our trouble are so largely religious, the remedy must also be essentially religious, whether we call it that or not. We must rethink and refeel our nature and destiny."[20]

The accuracy of White's analysis has been widely discussed and often challenged. Critics include John Passmore, a well-known philosopher of the

[17] L. White, "The Historical Roots of Our Ecological Crisis," *Science*, 155, no. 3767 (1967), 1205.
[18] White, "The Historical Roots of Our Ecological Crisis," 1206.
[19] White, "The Historical Roots of Our Ecological Crisis," 1207.
[20] White, "The Historical Roots of Our Ecological Crisis," 1207.

history of ideas, who argued in *Man's Responsibility for Nature* that White oversimplified by not recognizing that the Western negative attitude to nature originated more in Greek sources of Christian belief than in Hebrew sources.[21] Indeed, Greek ideas influenced Christian thinking, often in heretical directions, including early church heresies such as Docetism and Gnosticism, which were based on a devaluation of matter and the body. In his book *The Travail of Nature: The Ambiguous Ecological Promise of Christian Theology*, Paul Santmire agrees that Platonic ideas shaped a mistaken otherworldly idea of spirituality in Christian history, overly emphasizing the transcendence of God from the world, and even of humanity from nature, all of which grounded lack of concern if not outright hostility toward nature. However, he does identify and elaborate another theological ecological motif that places greater emphasis on God's immanence in creation and, consequently, on the intimate interrelationship between God, humans, and the wider creation. This motif places emphasis, as Santmire says, on "the human spirit's rootedness in the world of nature and on the desire of self-consciously embodied selves to celebrate God's presence in, with, and under the whole biophysical order."[22]

Particularly fascinating is Peter Harrison's partial agreement with White that the biblical account of creation played a role in an exploitative attitude toward nature. However, Harrison further clarifies that this attitude was linked to the idea of a fall, a straying of humans and nature from the divine plan. Therefore, the kind of positive domination over nature that was meant for humans is not realized in the current fallen world. Many traditional interpretations held that animal savagery, infertility of the ground, and pests and weeds were a consequence of Adam's fall. As Saint Paul proclaimed, "the creation groans in travail" for its eventual healing and restoration.[23] In this vein, many seventeenth-century advocates of early science thought that they could "improve" nature or partially restore it, and quite a few thought that subjugation of the earth meant simply to make it fit for agriculture.[24] In continuing his more nuanced analysis, Harrison mentions a handful of

---

[21] J. Passmore, *Man's Responsibility for Nature: Ecological Problems and Western Traditions* (New York: Scribner, 1974), 3–27, esp. 17.

[22] H. P. Santmire, *The Travail of Nature: The Ambiguous Ecological Promise of Christian Theology* (Minneapolis, MN: Fortress Press), 9.

[23] Romans 8:22.

[24] P. Harrison, "Subduing the Earth: Genesis 1, Early Modern Science, and the Exploitation of Nature," *The Journal of Religion*, 79, no. 1 (1999), 86–109.

obvious examples of this thinking, beginning with poet Thomas Traherne, who observed that the earth "had been a Wilderness overgrown with Thorns, and Wild Beasts, and Serpents: Which now by the Labor of many hands, is reduced to the Beauty and Order of *Eden*."[25] Politician John Pettus made explicit reference to "subduing the earth" and "conquering nature" aimed at "the replenishment of the first creation."[26]

The great poet John Donne encapsulated the mission of humanity that these other writers envisioned: "To rectify nature to what she was."[27] Therefore, in all fairness, we might say that the motivations of some early modern thinkers were aimed at restoring what they saw as a proper – but lost – dominion through a redemptive project. Of course, biblically and theologically, short of final eschatological redemption of all things, the human condition remains a mix of good and evil such that our activities in the natural environment are also mixed, some helpful and some harmful. Thus, the claim that Christian ideas categorically entail a despotic view of humanity over nature lacks nuance and perspective – and it invites further consideration of Christian thinking that assigns humanity a stewardship role.

Theological Reflections on Stewardship

Although White's analysis was simplistic, it served as a stimulus for a flurry of activity within the Christian community to articulate a positive role for humanity in the environment. Various sources of support have been brought into the discussion, most centrally to explain how humanity has responsibility for creation stewardship, either linking stewardship with dominion or eliminating talk of dominion. Reactions to this project range from endorsing its necessity to declaring its irrelevance to warning of its potential destructiveness. In fact, a creation stewardship rationale for earth care has received mixed reception among environmental movements, often depending on the domain of the environment at issue. For instance, to date theological articulation has been more welcome in discussions of species, habitat, and

[25] T. Traherne, *Christian Ethicks* (London: Printed for Jonathan Edwin, 1675), 103.

[26] J. Pettus, *Volatiles from the History of Adam and Eve* (London: Printed for T. Bassett, 1674), 83.

[27] J. Donne, *To Sir Edward Herbert, at Juliers in John Donne: The Major Works* (Oxford: Oxford University Press, 2000), 201, lines 33–34.

resource conservation than in discussions of animal rights issues. In this section, we review some of the traditional sources for Christian stewardship ideas as well as the more recent constructive Christian ecotheologies.

A great deal of exegetical work on Genesis argues that humanity is to care for creation rather than to exploit it. Major studies of the "dominion" language in Genesis 1 conclude that it implies that humanity has the vocation of caretaker or steward. Reinforcing this strategy, work on Genesis 2 links the language of humanity being created "in the image of God" with earth care, because a good God would care for his creation and would do so via human agency. Indeed, some scholars observe that in the imagery of Genesis 2:7, God forms the human (Hebrew: ha-adam) from the ground (Hebrew: ha-adamah), meaning that the human is explicitly an "earth-creature." A clear theme is the intimate connectedness of humanity with everything else that depends upon the earthly habitat. This is hardly a picture of humanity as aloof and domineering.[28]

Additional Old Testament sources for stewardship ideas include the texts where the Hebrew prophets urged a Sabbath for the land and connected dominion with caretaking.[29] Psalm 104 praises God for how creation displays his greatness, early verses declaring that the light is like his garment, that he stretched out the heavens like a tent, and that he set the earth on its foundations. Later verses of that Psalm then form a litany of things created by the Lord:

> He waters the mountains from his upper chambers;
>> the land is satisfied by the fruit of his work.
> He makes grass grow for the cattle,
>> and plants for people to cultivate –
>> bringing forth food from the earth: ...
> The trees of the Lord are well watered,
>> the cedars of Lebanon that he planted.
> There the birds make their nests;
>> the stork has its home in the junipers.
> The high mountains belong to the wild goats;

[28] R. Bauckham, *The Bible and Ecology*, 11–12; N. C. Habel, "Geophany: The Earth Story in Genesis 1," in N. C. Habel and S. Wurst (eds.), *The Earth Story in Genesis* (Sheffield: Sheffield Academic Press, 2000), 34–38; Claus Westermann, *Genesis*, trans. David E. Orton (Edinburgh: T. & T. Clark, 1988), 11.

[29] Leviticus 25:1–7; Exodus 23:10–11; Nehemiah 10:31; 2 Chronicles 36:20–21.

> the crags are a refuge for the hyrax . . . .
> How many are your works, Lord!
>   In wisdom you made them all;
>   the earth is full of your creatures.[30]

The last verse here – stating that creation is through wisdom – has been the basis for a good deal of scholarship related to theology, science, and the environment.

Celia Deane-Drummond is a prominent theologian who explores wisdom as a major thread running throughout both the Old and New Testaments. In *Creation through Wisdom: Theology and the New Biology*, she seeks to "give theology a clearer voice in the debate" between theology and the new biological sciences.[31] Her analysis of the New Testament on this subject finds once again that wisdom is linked to God, and indeed that Christ is identified in the Gospel of John as the wisdom of God, the *logos*.[32] This cosmic wisdom – or cosmic Christ – is the all-encompassing ground of order and unity in creation. Paul says in Colossians, "Through him all things were created . . . and in him all things hold together."[33] The biblical concept of wisdom, then, anchors a strong sense of God's immanence within creation, which counterbalances overemphasis on his transcendence and contributes to deeper reflection on creation in a world of science and technology. Such ideas as these contribute to a more holistic approach to knowledge and life under a theological construct.

In addition to the concept of divine immanence in creation, the New Testament contains many parables about stewards, although none depict humanity as caring for the earth per se. In Chapter 12 of the Gospel of Luke, Jesus tells a parable of a rich person appointing a "manager" over his possessions and servants during his absence and then later asking who is the faithful manager – the one who responsibly feeds the other servants or the one who assumed that his master would not return for a long time and began mistreating and beating the other servants. Clearly, the good manager treated all under his charge well, while the bad manager treated them harshly.

---

[30] Psalm 104:13–14, 16–18, 24.
[31] C. Deane-Drummond, *Creation through Wisdom: Theology and the New Biology* (Edinburgh: T&T Clark, 2000), xiv.
[32] Deane-Drummond, *Creation through Wisdom*, 49.   [33] Colossians 1:16–17.

Moving from biblical texts to the more systematic ideas of historic Christian theology, we find more support for the intrinsic value of nature. God in Genesis pronounces every physical thing he creates "good," bestowing upon it a value and place in the order of the world.[34] God embeds the rational creation – which bears the image of himself – in the material creation as a psychophysical being. Moreover, the Trinitarian God, in the Second Person, the Son, becomes identified with a particular physical human being, Jesus of first-century Nazareth. Thus, theologians assert that the Incarnation entails that God thought it no offense to come into his material creation. Moreover, the created world is treated not simply as having a present value but also as having a positive future value in the ultimate fulfillment of all things. Although popular versions of Christianity speak as if the material creation will be discarded, numerous theologians say that classical Christian eschatology envisions that the material creation will be in some sense restored and transformed. Indeed, some theologians argue that the doctrines of the bodily resurrection and ascension of Jesus entail that materiality has now been taken into the Godhead itself. For theologians reflecting on the environment, the fundamental point of such themes is that the physical creation has a high value that justifies responsible care for it.

As contemporary Christian thinkers have built into their theological work a significant dimension on nature and the environment, they have given rise to a movement called "ecotheology." The work of Christian theologian Jürgen Moltmann in this area is exemplary, and his book *God in Creation* gave major impetus to the effort of developing a theology of creation that engages ecological concerns besetting Western culture. A major theme of his work is social Trinitarianism – the idea that God is intrinsically a social, relational being – which then suggests a God who is relationally interactive with creation, both humanity and nature. Additional relational concepts of interconnectedness and mutuality also readily engage ecological interests. A related theme is Moltmann's emphasis on hope in the eschatological future, including the future of nature, because of God's loving sovereignty.[35]

Constructive Christian ecotheologies fall along a considerable range, some relying on classically orthodox Christian ideas and others relying on alternative philosophical frameworks and cosmologies. To help sort out these

---

[34] Genesis 1:4, 10, 12, 18, 21, 25, 31.
[35] J. Moltmann, *God in Creation* (1985; repr. Minneapolis, MN: Fortress Press, 1993).

ecotheologies, Paul Santmire has offered a helpful typology,[36] from which we draw two illustrative examples that are not classically orthodox in their approach. With a foundation in Alfred North Whitehead's process metaphysics, which stresses interdependence of all parts of a whole, John Cobb published *Is It Too Late? A Theology of Ecology* in 1971.[37] Rosemary Radford Ruether, feminist scholar and Catholic theologian, has worked energetically on ecofeminist theology in an effort that brings together three major concerns: ecology, feminism, and global justice.[38] Her strong opposition to God's transcendence – which she links to illegitimate elite oppression of weaker classes – leads her to describe God essentially as a cosmic force, the "material spiritual power for the renewal of life."[39]

By the same token, the other monotheistic traditions have their own ecotheologies. Jewish philosophers Abraham Joshua Heschel and Martin Buber, whose works stress inwardness and relationality, have inspired Jewish thought about ecology and influenced Christian thought as well.[40] Some Muslim theologians are also revisiting their theological foundations in order to develop authentic interpretations of Islam that are in tune with contemporary environmental concerns. In his piece "Toward an Islamic Ecotheology," K. L. Afrasiabi charts out this task.[41] In all such efforts, we see honest wrestling with common monotheistic themes – such as the ontology of God's relation to creation, the special place of humans, and the value of the nonhuman world. But the common aim is to associate God closely with nature, eliminate stereotypical monarchical views of humanity, and reinforce the importance of environmental care. The net result in Jewish and Islamic ecotheology, as in Christian ecotheology, is to advance a stewardship theme with an eye toward sustainability.

[36] H. P. Santmire, *Nature Reborn: The Ecological and Cosmic Promise of Christian Theology* (Minneapolis, MN: Fortress Press, 2000).

[37] J. B. Cobb, *Is It Too Late? A Theology of Ecology* (Beverly Hills, CA: Bruce, 1972).

[38] R. R. Ruether, *Gaia and God: An Ecofeminist Theology of Earth Healing* (San Francisco: HarperSanFrancisco, 1994).

[39] R. R. Ruether, "Ecofeminism and the Globalization," presentation at Garret-Evangelical Theological Seminary, May 20, 2015. Available online at www.garrett.edu/news-and-media/video-archive.

[40] D. M. Seidenberg, *Kabbalah and Ecology: God's Image in the More-Than-Human World* (New York: Cambridge University Press, 2015); H. Tirosh-Samuelson (ed.), *Judaism and Ecology: Created World and Revealed Word* (Cambridge, MA: Harvard University Press, 2002).

[41] K. L. Afrasiabi, "Toward an Islamic Ecotheology," *Hamdard Islamicus*, 18, no. 1 (1995), 33–44.

Theological understanding of stewardship can guide dedicated work on environmental concerns by individuals and organizations. A good example of an environmentally dedicated individual is found in Katherine Hayhoe, an evangelical Christian believer and one of the top climate scientists in the United States. Hayhoe's June 2019 testimony before the US House Committee on the Budget contained warnings about greenhouse gases, which she indicates are caused primarily by the use of fossil fuels.[42] In some venues, she has also warned against "climate fear," which is "turning into a new religion," with overly strict rules, guilt, and judgment enforced with all the "zeal of the Spanish Inquisition."[43] The Jewish Climate Initiative is a good example of a religious organization dedicated to the environment. This nonprofit entity – which was founded for the purpose of "Repairing the World" – sponsors interfaith environmental conferences that help encourage seminaries in North America, Rome, and the Holy Land to focus on ecological values that will eventually spread through communities and congregations.[44]

Science, Spirituality, and the Environment

Ecotheologians argue against undue emphasis on divine and human transcendence of nature, while seeking to develop understandings of God's immanence in and humanity's intimate relation to nature. Their theories of God generally relinquish more traditional ideas of any transcendence/immanence balance and instead accent immanence to the exclusion of transcendence. These theories vary a bit on the place of humanity, some assigning humanity a stewardship role based on inherent specialness and others immersing humans entirely in the natural world as part of a larger whole. Within the Christian tradition, Thomas Berry, a cultural historian and religion scholar, pioneered creation-centered theology, making the following public statement at an environmental conference in 1987:

---

[42] K. Hayhoe statement to United States House Committee on the Budget, *The Costs of Climate Change: Risks to the U.S. Economy and the Federal Budget*, Hearings on HR 109605, 116th Cong. June 11, 2019.

[43] K. Hayhoe in B. Ward, "The High and Low Points for Climate Change in 2019," Yale Climate Connections, December 11, 2019, www.yaleclimateconnections.org/2019/12/the-high-and-low-points-for-climate-change-in-2019/.

[44] For more information, see www.interfaithsustain.com/jewish-climate-initiative/.

"The universe is a communion of subjects, not a collection of objects. And listen to this: The human is derivative. The planet is primary."[45]

Rachel Carson, an American biologist and ecologist before ecology was officially a science, provides an interesting case study of nature-centered spirituality. Carson was a "religious humanist," whose life, work, and legacy inspired the modern environmental movement. Growing up in a Calvinist Presbyterian home with strong views of divine sovereignty, she, nevertheless, acquired from her mother a deep sense of the divine in nature that was influenced by the New England Transcendentalist tradition as much as by Christian thought. She saw a "wholeness" in all of life,[46] as did Henry David Thoreau, and she kept a copy of *Walden* at her bedside. In 1962, Carson's book *Silent Spring* appeared, warning about the indiscriminate use of synthetic pesticides and other harmful chemicals to the biosphere.[47]

In 1963, Carson gave lengthy testimony to the Senate subcommittee of the Committee on Government Operations, opening with a powerful statement:

> The contamination of the environment with harmful substances is one of the major problems of modern life. The world of air and water and soil supports not only the hundreds of thousands of species of animals and plants, it supports man himself . . . . Now we are receiving sharp reminders that our heedless and destructive acts enter into the vast cycles of the earth and in time return to bring hazard to ourselves.[48]

Because of her tireless efforts, Carson was awarded the Albert Schweitzer Medal of the Animal Welfare Institute, declaring in her acceptance speech, "Dr. Schweitzer has told us that we are not being truly civilized if we concern ourselves only with the relation of man to men. What is important is the relation of man to all life."[49]

---

[45] T. Berry at the North American Conference on Christianity and Ecology, August 1987, Indiana's Lake Webster Center. See also T. Berry and B. Swimme, "The Ecozoic Era," *Anima*, 20, no. 2 (1994), 105–118.

[46] See C. Lasher, "The Religious Humanism of Rachel Carson: On the 50th Anniversary of the Publication of Silent Spring," *Journal of Oriental Studies*, 22 (2012), 193–205.

[47] R. Carson, *Silent Spring* (Boston, MA: Houghton Mifflin Company, 1962).

[48] R. Carson to the Subcommittee on Reorganization and International Organizations of the Committee of Governmental Operations, *Environmental Hazards Control of Pesticides and Other Chemical Poisons*, Hearing on S Res. 27, 88th Cong., June 4, 1963.

[49] F. Stewart, "Small Winged Forms above the Sea: The Life of Rachel Carson," *Orion*, 14 (Winter 1995), 14–18, 18.

The focus on divine immanence is often linked to forms of spirituality based on a strong sense of kinship – and even a mystical bond – with the whole of nature. Seeking resources to support a spiritual vision of nature – or a "creation spirituality"[50] – some Christian thinkers have looked not only to the medieval mystics but also to pagan spiritualities. A prime example is Gaia in Greek mythology, the primal Mother Earth goddess, whose name is now a label for a new earth-centered holistic ecological philosophy, which is adaptable to various religions. Talk of Gaia was actually renewed on the science side in 1972 by atmospheric chemist and environmentalist James Lovelock, who published an article introducing the "Gaia hypothesis." The scientific hypothesis states that the earth is a living, complex entity, itself a living organism, that regulates its own temperature and chemistry to maintain conditions suitable for life. The system involves the biosphere, the atmosphere, the pedisphere, and the hydrospheres in an interrelated, self-regulating system.[51] A cybernetic feedback system among the biota – that is, among the total collection of organisms on the planet – leads to the homeostasis that allows habitability. Lovelock's 1988 book, *Ages of Gaia: A Biography of Our Living Earth*, presented his evidence for this, tracing the evolution of the world from early thermoacidophilic and methanogenic bacteria toward the oxygen-rich atmosphere that supports more complex life today.[52]

The Gaia hypothesis (Gaia theory, Gaia principle) – essentially, that the Earth is a living organism – was predictably controversial. Proponents heralded Gaia as having the potential for a science-inspired cultural transition. W. D. Hamilton called Gaia downright Copernican, altering the human self-perception of centrality on Earth just like heliocentric theory altered our perception of centrality of Earth in the local solar system. Critics initially rejected the Gaia hypothesis for being teleological in purporting that the Earth itself has the goal of being a place for life. Dawkins argued that the Earth cannot be a living thing because life is produced by natural selection between individuals seeking reproductive success. However, as microbiologist Lynn Margulis collaborated with Lovelock, they explicitly tried to bring

[50] M. Fox, *Creation Spirituality: Liberating Gifts for the Peoples of the Earth* (San Francisco HarperSanFrancisco, 1991).
[51] J. E. Lovelock, "Gaia as Seen through the Atmosphere," *Atmospheric Environment*, 6, no. 8 (1972), 579–580.
[52] J. Lovelock, *The Ages of Gaia: A Biography of Our Living Earth* (Oxford: Oxford University Press, 1988).

the Gaia hypothesis into greater consonance with evolutionary principles. Margulis stated that "Darwin's grand vision was not wrong, only incomplete." She explains as follows:

> In accentuating the direct competition between individuals for resources as the primary selection mechanism, Darwin (and especially his followers) created the impression that the environment was simply a static arena.[53]

Further developments in the Gaia hypothesis attempted to align it with theories and findings in Earth system science, biogeochemistry, and systems ecology.

Nonetheless, scientists generally remained resistant to Lovelock's hypothesis. Stephen Jay Gould insisted that it was "a metaphor, not a mechanism." John Maynard Smith called it "an evil religion," mere pseudoscience, with an idealistic, quasi-religious dimension that many of its lay adherents identified with. In fact, public reaction to Gaia was overwhelmingly positive. The idea that our planet was somehow alive also resonated with poets, writers, pagans, and even some churchgoers, and environmentalists energetically embraced the idea. In light of the mixed reception of Gaia, philosopher of evolutionary biology Michael Ruse has thoroughly explored what we might call our "love–hate" relationship with Gaia.[54]

Early on, Lovelock and Margulis produced a series of papers arguing for Gaia and particularly focusing on Earth's temperature during its history. Whereas increasing sun temperatures would predict a resulting rise in Earth's temperature, Earth's temperature has remained relatively stable. Now, a key feature of life is homeostasis – that is, the ability to maintain a stable balance through dynamic interacting processes. They claim that the fact that the temperature of Earth has not risen with the rising temperature of the sun, which is a giant thermonuclear reactor, displays Earth's ability to maintain homeostasis. Among the reasons for this temperature stability is the planet's entire biota, which changes the composition of the gases on Earth in a way that moderates the effects of the sun's heat. In effect, the planet is regulating its own "greenhouse effect" by modulating the rate at which trapped solar heat is released into space. We cannot pursue the details

---

[53] L. Margulis and G. Hinkle, "The Biota and Gaia: One Hundred Fifty Years of Support for Environmental Sciences," in L. Margulis and D. Sagan (eds.), *Slanted Truths: Essays on Gaia, Symbiosis, and Evolution* (New York: Copernicus, 1997), 214.

[54] M. Ruse, *The Gaia Hypothesis* (Chicago: University of Chicago Press, 2013), 25–42.

of the science here but note that the idea is that Earth adjusts its gaseous mantle to accommodate temperature changes just as a human perspires when temperature changes. In doing so, Earth's feedback mechanisms maintain an environment conducive to its biota.

Over time, reactions to the Gaia hypothesis within the scientific community have varied – from ignoring it, to ridiculing it, to using it as a fruitful paradigm. Particularly, studies in the multidisciplinary fields of Earth system science and biogeochemistry treat it as a testable theory that provides a number of useful predictions, such as in regard to climate.[55] To be precise, the original hypothesis was proved technically mistaken as it was discovered that it is not life alone but the whole Earth system that self-regulates. In 2001, the European Geophysical Union declared, "The Earth System behaves as a single, self-regulating system comprised of physical, chemical, biological, and human components."[56]

Reactions to Gaia in the larger public have been broadly positive, perhaps largely because Gaia thinking is open textured and hence able to be incorporated into various perspectives – from traditional religious understandings of Earth as a divine creation worthy of awe and respect to neopagan views that Earth is our spiritual source and can be encountered mystically to secular visions of an interconnected planet that we all must learn to sustain. Whatever the ultimate status of the Gaia hypothesis turns out to be, as science or as religion, it has certainly stimulated helpful and productive discussion and research.

Secular Perspectives on Ecological Concerns

Over the past several decades, it has not been just religious groups that have taken an interest in the environment. Secular groups have been cultural players in environmental concerns as well, also seeking greater understanding of ecology and paths to sustainability generated by their own nonreligious outlooks. Although some nonbeliever environmentalists are happy for any support on the issues from any quarter, others are extremely resistant to

---

[55] For experimental support see J. Lovelock, "Hands Up for the Gaia hypothesis," *Nature*, 344 (1990), 100–102. For predictions see T. Volk, *Gaia's Body: Toward a Physiology of Earth* (Cambridge, MA: MIT Press, 2003).

[56] Challenges of a Changing Earth: Global Change Open Science Conference Amsterdam, The Netherlands, July 13, 2001. Available online at www.igbp.net.

religious involvement in our political life. Yet, Roger Gottlieb, a scholar in religion and ecology, offers a caveat, stating that "any unbiased look at the last century of political life shows that society is endangered as much by fanaticism of the secular variety as it is by that of the faithful."[57] Some religious advocates argue that ideals of democracy are rooted in Judeo-Christian ideas that all persons are created equal under God, and observe that many great causes, including abolition and suffrage, were strongly driven by the faith of passionate believers.[58] Besides, a more pragmatic approach necessarily includes environmentalists of all stripes to build alliances and make compromises where values intersect.

Nonetheless, many believe that a purely secular outlook is sufficient to support environmentalist thought and action. Naturalism or materialism form the typical worldview of much religious nonbelief, but it is very difficult for such views at the metaethical level to generate intrinsic value for nature or to propose anything other than a prudential ethic toward nature. Most secular thinkers, then, argue for the utilitarian value of environmentalist norms and ethical agendas, many of which may resemble those of religious environmentalists. Couched in traditional Darwinian terms, the argument is that humans are part of the whole natural world, materially and biologically, and thus must be pragmatically concerned for its welfare. Hence, the Darwinian naturalist faces the issue of whether some kind of value can be derived from amoral nature, something Homes Rolston III tries to do in his own way:

> Environmental ethics ... is the most altruistic, global, generous, comprehensive ethic of all, demanding the most expansive capacity to see others ... it is naturalized ethics in the comprehensive sense, humans acting out of moral conviction for the benefit of nonhuman others.[59]

---

[57] R. Gottlieb, "Religion and Ecology – What Is the Connection and Why Does It Matter?," in R. Gottlieb (ed.), *The Oxford Handbook of Religion and Ecology* (New York: Oxford University Press, 2006), 10.

[58] For a few examples, see the following: R. Anstey, "Slavery and the Protestant Ethic," *Historical Reflections*, 6, no. 1 (1979), 157–181; J. B. Stewart, *Holy Warriors: The Abolitionists and American Slavery*, rev. edn. (New York: Macmillan, 1997); F. Dudden, *Fighting Chance: The Struggle Over Woman Suffrage and Black Suffrage in Reconstruction America* (New York: Oxford University Press, 2011); John Wesley, *Thoughts upon Slavery* (Pamphlet, 1774).

[59] H. Rolston III, *Genes, Genesis, and God: Values and Their Origins in Natural and Human History* (New York: Cambridge University Press, 1999), 288.

For Rolston's scientific naturalism, nature and the care of nature are the highest value.

Even if such arguments do not quite establish nature itself as a sufficient ground for environmental values and ethics, secular thinkers essentially appeal to the value of "enlightened self-interest." In explaining evolutionary ethics, biologist David Sloan Wilson and philosopher Elliot Sober use the analogy of a group of people caught in a lifeboat, bound together by "the prospect of a common fate." They can survive only if they all work together – and in this case, Earth is the lifeboat.[60] The American Humanist Association stressed our mutual dependence on one another and the environment in its 2019 resolution:

> Humanity's global interdependence is central to the humanist ethic. That interdependence extends far beyond humans and includes the other living beings with whom we share the planet, the water we drink, the air we breathe, and the lands we inhabit. Just as we depend on this planet to sustain us, Earth's ecosystems depend on us to be good stewards and take responsibility for the impact human activity has on our shared environment. Humanists and all others should make conservation, ecological education, and environmental stewardship top priorities to safeguard against all forms of environmental degradation and ensure a healthy planet.[61]

Believing that this is the only life we have, atheist values reflect a secular version of humanism that emphasizes human effort in solving our problems, including environmental problems. Some secular humanists even argue that their humanism is a stronger support for environmental ethics than religion because they are not looking for divine assistance. The Humanists UK society promotes a similar view.[62]

Religion, Politics, and Environmental Issues

Religious perspective and political ideology have long influenced attitudes toward environmental issues, but recent academic research has probed more

---

[60] E. Sober and D. S. Wilson, *Unto Others: The Evolution and Psychology of Unselfish Behavior* (Cambridge, MA: Harvard University, 1998), 334–336.

[61] American Humanist Association, "Resolution on the Environment," adopted December 7, 2019, Washington, DC. Available online at https://americanhumanist.org/key-issues/statements-and-resolutions/resolution-on-environmentalism/.

[62] For more information on the Humanists UK society, see https://humanism.org.uk.

deeply into their interaction. A study by political scientist Matthew Arbuckle analyzed the complexities of how religious views across the conservative–liberal spectrum intersect political environmental ideology in the United States. Focusing on the controversial issue of climate change, Arbuckle found no neat correlations but noted, perhaps counterintuitively, that political liberals are more influenced by their religious positions.[63] Similarly, environmental psychologists Aimie L. B. Hope and Christopher R. Jones have studied religious attitudes in the United Kingdom toward the issue of climate change and carbon dioxide (CO_2) reduction methods in particular. One fascinating finding was that religious believers with the hope of an afterlife have less urgency and anxiety about climate change matters than their secular counterparts.[64]

Interestingly, various surveys conducted at different times by the Pew Research Center in the United States found no significant correlations between religion and political views on the environment. This includes almost all of the well-known concerns – climate change, deforestation, building more nuclear power plants, hydraulic fracturing, offshore drilling, species extinction, and so forth.[65] In fact, on most environmental issues, Pew found no great divergence among religiously affiliated and religiously unaffiliated groups and no obvious pattern or predictor of how they might agree or disagree on particular issues. Such a result strongly suggests that the greater influence is political party, education, and other factors on environmental outlook.[66] However, some studies (not Pew) detect more correlation between religion and higher environmental concern when the religious perspective includes themes of stewardship and duty to the present as well as a restrained emphasis on an afterlife.[67] In this vein, the Evangelical Environmental Network, founded in 1993, would be one example of a

---

[63] M. B. Arbuckle, "The Interaction of Religion, Political Ideology, and Concern about Climate Change in the United States," *Society & Natural Resources*, 30, no. 2 (2017), 177–194.

[64] A. L. B. Hope and C. R. Jones, "The Impact of Religious Faith on Attitudes to Environmental Issues and Carbon Capture and Storage (CCS) Technologies," *Technology in Society*, 38 (2014), 48–59.

[65] Pew Research Center, "Religion and Science," October 22, 2015.

[66] See Pew Research Center, "Catholics Divided Over Global Warming," June 16, 2015, on the role of religious affiliation on beliefs about climate change.

[67] K. K. Wilkinson, *Between God & Green: How Evangelicals Are Cultivating a Middle Ground on Climate Change* (Oxford: Oxford University Press, 2012).

Christian ministry organization aimed at reducing environmental degradation based on the idea of human responsibility to care for God's creation.[68]

Probably no Christian leader has gained as much worldwide celebrity for engaging environmental issues as Pope Francis. He has challenged what he perceives as hypercapitalistic patterns of consumption and endorsed environmental protection. He has expressed interest in a broad range of issues, including pollution, clean drinking water, biodiversity, vanishing ecosystems, and social inequality, all with the flavor of his Argentinian liberation theological outlook. His encyclical *Laudato si'* (released in June 2015, shortly before the international Conference of Parties Paris agreement on climate change) charted an overarching vision for the role of Christians in "caring for our common home."[69] This encyclical opens with the *Canticle of the Creatures* from St. Francis of Assisi, whose name the pope took as an inspiration and guide: "Praise be to you, my Lord, through our Sister, Mother Earth, who sustains and governs us, and who produces various fruit with colored flowers and herbs."[70]

As the encyclical proceeds, it methodically reviews past papal encyclicals, calling for environmental responsibility and then mounting a lengthy argument for theological transformation and spiritual renewal centered on ecological awareness. Putting ecotheology prominently on the theological map for the world to see, Pope Francis's basic rational is twofold: that the doctrine of creation entails that human beings, who are all created in God's image, whether or not they believe in God, should care for the Earth as our common home, and that, additionally, the doctrines of grace and redemption particularly justify Christian involvement in caring for Earth. On the issue of climate change, he sides with the view that it is humanly caused, largely due to the production of greenhouse gases since the Industrial Revolution, and he calls for measures to mitigate and reduce the gases.

In tone and substance, Pope Francis is acting not just as priest but as prophet as well, urging us to heed our moral duties to the environment and warning of undesirable consequences if we do not. Given that both religion

---

[68] For more information about the Evangelical Environmental Network, see https://creationcare.org.

[69] Francis, encyclical *Laudato Si'* (2015). Available online at www.vatican.va.

[70] Francis of Assisi, "Canticle of the Creatures," in R. J. Armstrong, J. A. W. Hellmann, and W. J. Short (eds.), *Francis of Assisi: Early Documents* (Hyde Park, NY: New York City Press, 1999), vol. I, 113–114.

and politics deal with ideas about how we should live together, the religious influence that this encyclical could exert on politics, through those who take it seriously, could be considerable. Pope Francis insightfully states that acceptance of a theological outlook does not always translate into sustained action and therefore calls for "an ecological spirituality grounded in the convictions of our faith"[71] that motivates us to passionate concern for the protection of our world. He closes the encyclical with a prayer: "O Lord, seize us with your power and your light, help us to protect all life, to prepare for a better future, for the coming of your kingdom of justice, peace, love and beauty: Praise be to you! Amen."

All major religions are looking to their theological and spiritual resources to meet the environmental challenge. However, given that religious believers differ over their assessment of environmental issues, we must press our survey a bit further. Theological teachings reject despotism over nature; humanity's positive relation to nature has been strongly affirmed; and serious religious believers endorse reasonable conservation of natural resources. Hence, many of the disagreements among believers actually appear to be political, regarding what policies and agendas are reasonable and acceptable.

In the United States, for example, believers who lean politically conservative generally caution against increasing the power of government. They readily identify waste and corruption when government mandates overly restrictive environmental regulations or uses public money to subsidize environmental causes. Conservatives also cite the massive and seemingly annual forest fires in California as a case of imbalanced government overreach. Although political liberals say that the destructive fires are a result of global warming, many conservatives argue that California's environmental regulations prevent standard forest husbandry procedures from clearing flammable underbrush, procedures successfully employed by other states with more realistic regulations to prevent forest fires.[72] Frequently

---

[71] Francis, *Laudato Si'*, 3.216.

[72] See the following: C. Fiorina, "The Man-Made Water Shortage in California," time.com, April 7, 2015, https://time.com/3774881/carly-fiorina-california-drought-environmentalists/; D. DeVore, "Wildfires Caused By Bad Environmental Policy Are Causing California Forests To Be Net CO_2 Emitters," forbes.com, February 25, 2019, www.forbes.com/sites/chuckdevore/2019/02/25/wildfires-caused-by-bad-environmental-policy-are-causing-california-forests-to-be-net-co2-emitters/.

mentioned on the long list of conservative complaints is the infamous Solyndra Solar project, which was funded with half a billion taxpayer dollars from President Barack Obama's green energy program. Although Solyndra promised cheaper solar solutions to energy, political conservatives argue that their business model was unworkable from the beginning, always requiring backup energy to solar and requiring government subsidy to operate. Conservatives also note that when Solyndra quickly went bankrupt, the owners still profited handsomely.[73]

Increasingly, the correlation between corruption and government-funded environmental projects is attracting academic study as well as popular attention. In 2016, a study in Italy revealed a statistically significant link between generous public policy for wind energy projects and "the formation of criminal associations between entrepreneurs and politicians able to influence the licensing process."[74] All this said, religious persons who are politically liberal or progressive still tend to favor extensive government involvement in the environment in the pursuit of policies they hope will ensure a sustainable future. Some admit that such efforts may be imperfect but insist that we must do something before we reach a point of no return. Some even bring environmental concern under the broader banner of concern for social justice.[75] In the end, the debate appears to be highly political rather than religious, involving differing visions of the power of the state in relation to the individual, business, and other organizations.

Although there are many more complexities at the intersection of religion, politics, and the environment than we can sort out here, it is still quite clear that religion must be engaged in our discussions of the environment. Religion may ultimately be one of the most important influences on our course of action. As Lynn White observed, "What people do about their ecology depends on what they think about themselves in relation to things around them. Human ecology is deeply conditioned by

---

[73] S. Dinan, "Obama clean energy loans leave taxpayers in $2.2 billion hole," *The Washington Times*, April 27, 2015, www.washingtontimes.com/news/2015/apr/27/obama-backed-green-energy-failures-leave-taxpayers/.

[74] C. Gennaioli and M. Tavoni, "Clean or Dirty Energy: Evidence of Corruption in the Renewable Energy Sector," *Public Choice*, 166 (2016), 261–290.

[75] A. J. Baugh, *God and the Green Divide: Religious Environmentalism in Black and White* (Oakland: University of California Press, 2017), 9.

beliefs about our nature and destiny – that is, by religion."[76] Religion as a social force can help shape how we respond as a world community to the environmental crisis, helping humanity learn how to live together, how to relate to other species and nature as a whole, and how to move in a positive direction.

[76] White, "The Historical Roots of Our Ecological Crisis," 1206.

Glossary

abiogenesis The origin of life from nonliving matter.

accommodationism In the science–religion debates, the attempt or
perceived attempt, on the part of either side, to find common ground.

adaptation Any heritable characteristic of an organism that improves its
ability to survive and reproduce in its environment.

agnosticism (agnostic) The position of not knowing whether or not
god exists.

anthropic argument A version of the teleological argument explaining the
physical laws that allow intelligent life by reference to the existence
of God.

antirealism The epistemological position that we cannot have knowledge of
some type of facts; the ontological position that there are no relevant
facts to know in some areas of life and the world.

atheism The denial that there is a god (see also agnosticism).

Cambrian period (Cambrian explosion, radiation) The relatively brief
evolutionary event beginning about 530 million years ago when a wide
variety of animals burst into the organic world as indicated by the
fossil record.

creation The theological doctrine asserting that a divine act brought the
universe and everything in it into existence; not to be conflated with
Creationism or Creation Science.

Creationism The religious fundamentalist (literalist) belief that the
scriptural narratives of divine creation of the world contain factually
correct information regarding the origin of the world and the instant
formation of all species.

defeater In contemporary epistemology, a belief that, if accepted, makes it
irrational to accept another belief that is in question.

deism The view that an omnipotent, omniscient creator brought the universe into existence and let it operate by natural laws without specific divine interventions or miracles.

design argument One type of teleological argument that reasons that the order and complexity in the world requires an intelligent being to establish it.

DNA Deoxyribonucleic acid, in a double-helix structure, which carries the genetic instructions used in the development, functioning, and reproduction of all living organisms (see also RNA).

emergence (emergentism) The process whereby larger or more complex patterns or powers occur through the interactions of smaller or simpler entities.

emergent dualism A view of the mind–body relationship holding that mind and mental properties arise from complex physical arrangements in the brain rather than being separate nonmaterial substance or being completely reducible to the physical brain.

eschatology The area of theology pertaining to the ultimate culmination of history and fulfillment of God's plans for creation and humanity.

ethology The scientific study of animal behavior, usually in the context of natural habitat and conditions.

evolution The change in heritable traits of biological populations over successive generations, giving rise to diversity in individuals and among species.

fine-tuning (argument) The argument that the precise balance of the fundamental physical constants of the universe are best explained by reference to an intelligent supreme being and not by chance.

fundamentalism A religious outlook based on belief in strict, literal interpretation of sacred scripture. Supports Creationism (see Creationism).

gene Most basically, the molecular unit of heredity that contains an organism's inherited phenotypic traits or, more technically, the locus of DNA that encodes a functional RNA product.

genome An organism's complete set of DNA, including all of its genes, which contain all information needed to build and maintain that organism.

genotype The collection of genes specific to an individual (see phenotype).

hominids Often known as the Great Apes, the large taxonomic family of primates that includes orangutans, gorillas, chimpanzees, and humans.

hominins The group consisting of all humans, modern and extinct, which are in turn a part of the larger family of primates known as hominids, which includes orangutans, gorillas, chimpanzees, and human beings.

Homo sapiens Homo is the genus, followed by *sapiens* as species, to form a term that literally means "wise man" or "knowing man."

imago Dei The theological term meaning "God's Image," which indicates the view that humans reflect or resemble God in certain important ways, usually in regard to rationality, morality, personhood, and will.

immanent The divine attribute of being present with the whole of creation while remaining ontologically distinct.

incarnation The Christian doctrine that God in the Second Person of the Trinity became one with the first-century person Jesus.

Intelligent Design (ID; argument) The position purporting to be a scientific theory in biology that certain kinds or levels of organic complexity (irreducible, specified) cannot have been produced by natural selection operating on chance variations and thus are evidence of a designing intelligence.

liberalism (liberal) The theological view that supernaturalism in scripture reflects a prescientific mentality, that literal interpretation of scripture is unnecessary, and that social action is religiously more important than individual salvation; often contrasted with fundamentalism.

literalism See fundamentalism.

materialism The philosophical view that everything that exists is material or physical in nature.

mechanistic explanation (mechanism) The type of explanation that cites a physical cause or causes as bringing about the phenomenon in question.

mentalism The view that mind is the fundamental reality at work in the universe.

metaethics The second-order discipline studying the grounds of ethical obligation, the structure of ethical arguments and reasons, and the meaning of ethical terms.

metaphysics (metaphysical) The branch of philosophy dealing with the nature and structure of ultimate reality.

metazoan A zoological group consisting of the multicellular animals, living or extinct, that have cells differentiated into tissues, organs, etc.

methodological naturalism (methodological) The philosophical position that science proceeds by seeking natural causes for natural phenomena and not recognizing supernatural causes as the basis for its explanations.

miracle Classically, a violation of a law of nature by a divine being, which is an event that otherwise would not have occurred in the regular course of nature.

moral argument A type of theistic argument that reasons that human morality is ultimately caused or explained by the existence of a moral God.

mutation In biology, a randomly occurring permanent change of the nucleotide sequence of the genome of an organism.

natural selection A key mechanism of evolution involving the differential survival and reproduction of individuals that differ in phenotype.

naturalism (metaphysical naturalism) The philosophical worldview resting on the metaphysical position that physical nature is the fundamental reality and that no deity exists or relates to the world, often linked to a strong empiricist epistemology.

neuroscience Considered a branch of biology, the scientific study of the nervous system, including the brain.

noetic Pertaining to believing and knowing.

normative ethics (substantive ethics) The branch of ethics dealing with actual (first-order) obligations, typically in terms of acts or rules.

ontology (ontological) The branch of philosophy, often associated with metaphysics, that deals with what has being or what kinds of things have being.

organism An individual living thing with interdependent parts making up one whole.

Paleolithic (age, era, period) The early prehistoric phase of the Stone Age, beginning about 2.5 million years ago and ending about 10,000 years ago.

paleontology The field of science that studies life and ecologies that existed in the distant past as based on fossils from the pertinent geological periods.

paradigm In a community of researchers, the shared assumptions, values, ideas, and problems that govern a normal period of scientific activity.

phenotype The observable traits of an individual (see genotype).

providence The theological concept that God somehow guides and interacts with the world for wise and good purposes.

random variation See mutation.

rationalism In the Enlightenment, the philosophy that either the content or the form of knowledge is produced by the mind.

realism The ontological position that there are objective realities outside the human mind; the epistemological position that our cognitive powers are adequate to know these objective realities.

reductionism Methodological reductionism is the procedure of describing a complex phenomenon in terms of simpler or constituent parts; metaphysical reductionism is the strong thesis that a complex phenomenon is merely the combination of its parts and cannot represent a higher-level reality.

RNA Ribonucleic acid, a polymeric molecule, which is involved in various biological roles in coding, decoding, regulation, and expression of genes (see also DNA).

skepticism (skeptic) The antirealist view, usually in epistemology but sometimes in moral theory, that we cannot know certain claims to be true, either because of the subject matter being beyond us or because our powers of knowing have deficiencies.

social Darwinism The philosophy of society, promulgated by Herbert Spencer and others, that the concepts of natural selection and survival of the fittest in biology can be applied to sociology, economics, and political theory in terms of a universal law of progress.

supervenience A concept with wide application in analytic philosophy to the effect that a set of properties A supervenes on another set B such that no two things can differ with respect to A-properties without also differing with respect to their B-properties.

teleological argument A type of theistic argument that cites the order and regularity of the world as requiring explanation by reference to a supremely intelligent being who orders it for an end or goal. (Design arguments are considered a subtype of this broad category of theistic argument.)

teleological explanation A type of explanation that refers to a purpose or intended end state, generally as the intention of an intelligent being.

teleology The quality of being end directed or goal directed, either inherently or by direction of another.

theism The belief in an omnipotent, omniscient, wholly good God who created, sustains, and interacts with the world.

theistic evolution The view that theism and evolution can be combined in a larger perspective.

Trinity In Christian theology, the doctrine that God is three divine Persons (Father, Son, Holy Spirit) in one being.

variation See mutation.

vitalism The view that life is an intangible element distinct from the physical.

worldview A comprehensive philosophical explanation of reality, knowledge, morality, and humanity that has implications for all major areas of life and the world.

Further Reading

Alexander, D. R., and R. Numbers (eds.), *Biology and Ideology from Descartes to Dawkins* (Chicago: University of Chicago Press, 2010).

Ayala, F. J., and R. Arp (eds.), *Contemporary Debates in Philosophy of Biology* (Malden, MA: Wiley-Blackwell, 2010).

Ayala, F. J., *Darwin's Gift to Science and Religion* (Washington, DC: Joseph Henry Press, 2007).

Barbour, I. G., *Religion and Science: Historical and Contemporary Issues* (New York: HarperOne, 1997).

Barnhill, D. L., and R. S. Gottlieb (eds.), *Deep Ecology and World Religions: New Essays on Sacred Grounds* (Albany, NY: State University Press of New York, 2001).

Bedau, M. A., and P. Humphreys, *Emergence: Contemporary Readings in Philosophy and Science* (Cambridge, MA: A Bradford Book, 2008).

Boyer, Pascal, *Minds Make Societies: How Cognition Explains the World Humans Create* (New Haven: Yale University Press, 2018).

Carroll, S. B., J. K. Grenier, and S. D. Weatherbee, *From DNA to Diversity: Molecular Genetics and the Evolution of Animal Design*, 2nd edn. (Malden, MA: Blackwell, 2005).

Collins, F., *The Language of God: A Scientist Presents Evidence for Belief* (New York: Free Press, 2006).

Cooper, D. E., and J. A. Palmer (eds.), *Spirit of the Environment: Religion, Value and Environmental Concern* (London: Routledge, 1998).

Coyne, J. A., *Faith Versus Fact: Why Science and Religion Are Incompatible* (New York: Penguin Books, 2016).

Darwin, C., *On the Origin of Species* (London: Murray, 1859). Find online at: http://darwin-online.org.uk and www.darwinproject.ac.uk.

Dawkins, R., *The Blind Watchmaker* (New York: W. W. Noton, 1986).

The Selfish Gene (Oxford: Oxford University Press, 1976).

de Waal, F., P. S. Churchland, T. Pievani, and S. Parmigiani (eds.), *Evolved Morality: The Biology and Philosophy of Human Conscience* (Leiden: Brill, 2014).

Deane-Drummond, C., *Creation through Wisdom: Theology and the New Biology* (Edinburgh: T&T Clark, 2000).

Dennett, D. C., *Darwin's Dangerous Idea: Evolution and the Meanings of Life* (New York: Simon & Schuster, 1995).

Dennett, D. C., and A. Plantinga, *Science and Religion: Are They Compatible?* (New York: Oxford University Press, 2011).

Dupré, J., *Darwin's Legacy: What Evolution Means Today* (New York: Oxford University Press, 2003).

Edelmann, J., *Hindu Theology and Biology: The Bhagavat Purana and Contemporary Theory* (New York: Oxford University Press, 2012).

Foltz, R. C., *Worldviews, Religion, and the Environment: A Global Anthology* (Boston, MA: Cengage Learning, 2002).

Gottieb, R. S. (ed.), *Oxford Handbook of Religion and Ecology* (New York: Oxford University Press, 2006).

Joyce, R., *The Evolution of Morality* (Cambridge, MA: MIT Press, 2006).

Larson, E. J., *The Creation-Evolution Debate: Historical Perspectives* (Athens: University of Georgia Press, 2007).

Lineweaver, C. H., P. C. W. Davies, and M. Ruse (eds.), *Complexity and the Arrow of Time* (Cambridge: Cambridge University Press, 2013).

Mayr, E., *What Evolution Is* (New York: Basic Books, 2001).

McGrath, A., *Darwinism and the Divine: Evolutionary Thought and Natural Theology* (Malden, MA: Wiley-Blackwell, 2011).

Murray, M. J., *Nature Red in Tooth and Claw: Theism and the Problem of Animal Suffering* (Oxford: Oxford University Press, 2008).

Numbers, R., *The Creationists: From Scientific Creationism to Intelligent Design*, expanded edn. (Cambridge, MA: Harvard University Press, 2006).

Peacocke, A., *God and the New Biology* (London: J. M. Dent & Sons, 1986)

Peterson, M. L., and M. Ruse, *Science, Evolution, and Religion: A Debate about Atheism and Theism* (New York: Oxford University Press, 2016).

Pollack, R., *The Faith of Biology and the Biology of Faith: Order, Meaning, and Free Will in Modern Medical Science* (New York: Columbia University Press, 2000).

Post, S. G., L. G. Underwood, J. P. Schloss, and W. B. Hurlbut (eds.), *Altruism and Altruistic Love: Science, Philosophy, and Religion in Dialogue* (New York: Oxford University Press, 2002).

Rapport, M. B., and C. J. Corbally, *The Emergence of Religion in Human Evolution*, Routledge Studies in Neurotheology, Cognitive Science and Religion (London: Routledge, 2020).

Rosenberg, S. (ed.), *Finding Ourselves after Darwin: Conversations on the Image of God, Original Sin, and the Problem of Evil* (Grand Rapids: Baker Academic, 2018).

Ruse, M., *Can a Darwinian Be a Christian?* (Cambridge: Cambridge University Press, 2001).

  Science and Spirituality: Making Room for Faith in the Age of Science (Cambridge: Cambridge University Press, 2010).

Russell, R. J., W. R. Stoeger, and F. J. Ayala (eds.), *Evolutionary and Molecular Biology: Scientific Perspectives on Divine Action* (Notre Dame: University of Notre Dame Press, 2006).

Schloss, J. P., and M. J. Murray (eds.), *The Believing Primate* (Oxford: Oxford University Press, 2010).

Tanner, R., and C. Mitchell, *Religion and the Environment* (New York: Palgrave Macmillan, 2002).

Taylor, B. (ed.), *Encyclopedia of Religion and Nature*, 2 vols. (New York: Continuum, 2005).

Torrey, E. F., *Evolving Brains, Emerging Gods: Early Humans and the Origins of Religion* (New York: Columbia University Press, 2017).

van Huyssteen, J. W., *Alone in the World?: Human Uniqueness in Science and Theology* (Grand Rapids: Eerdmans, 2006).

Venema, D. R., and S. McKnight, *Adam and the Genome: Reading Scripture after Genetic Science* (Grand Rapids: Brazos Press, 2017).

Wilson, E. O., *Sociobiology*, 25th anniversary edn. (Cambridge, MA: Harvard University Press, 2000).

Index

Printed in the United States
by Baker & Taylor Publisher Services